Electronic Circuits: Fundamentals and Applications

Electronic Circuits: Fundamentals and Applications

Second edition

Michael Tooley, BA
Director of Learning Technology
Brooklands College

Newnes

OXFORD   AUCKLAND   BOSTON   JOHANNESBURG   MELBOURNE   NEW DELHI

Newnes
An imprint of Butterworth-Heinemann
Linacre House, Jordan Hill, Oxford OX2 8DP
225 Wildwood Avenue, Woburn, MA 01801-2041
A division of Reed Educational and Professional Publishing Ltd

 A member of the Reed Elsevier plc group

First published 1995 as *Electronic Circuits Student Handbook*
Reprinted 1999
Second edition 2002

© Michael Tooley 1995, 2002

British Library Cataloguing in Publication Data
A catalogue record for this book is available from the British Library

ISBN 0 7506 5394 9

Typeset by Integra Software Services Pvt. Ltd., Pondicherry 605 005, India
www.integra-india.com
Printed and bound in Great Britain

Contents

Preface

This book has been designed to help you understand how electronic circuits work. It will provide you with the basic underpinning knowledge necessary to appreciate the operation of a wide range of electronic circuits including amplifiers, logic circuits, power supplies and oscillators.

The book is ideal for people who are studying electronics for the first time *at any level* including a wide range of school and college courses. It is equally well suited to those who may be returning to study or who may be studying independently as well as those who may need a quick *refresher*.

The book has 14 chapters, each dealing with a particular topic, and seven appendices. The approach is topic-based rather than syllabus-based and each major topic looks at a particular application of electronics. The relevant theory is introduced on a progressive basis and delivered in manageable chunks.

In order to give you an appreciation of the solution of simple numerical problems related to the operation of basic circuits, worked examples have been liberally included within the text. In addition, a number of problems can be found at the end of each chapter and solutions are provided at the end of the book. You can use these end-of-chapter problems to check your understanding and also to give you some experience of the 'short answer' questions used in most in-course assessments. For good measure, we have included 70 revision problems in Appendix 2.

At the end of the book we have included 21 sample coursework assignments. These should give you plenty of 'food for thought' as well as offering you some scope for further experimentation. It is not envisaged that you should complete all of these assignments and a carefully chosen selection will normally suffice. If you are following a formal course, your teacher or lecturer will explain how these should be tackled and how they can contribute to your course assessment.

While the book assumes no previous knowledge of electronics you need to be able to manipulate basic formulae and understand some simple trigonometry in order to follow the numerical examples. A study of mathematics to GCSE level (or equivalent) will normally be adequate to

satisfy this requirement. However, for those who may need a refresher or have had previous problems with mathematics, Appendix 6 will provide you with the underpinning mathematical knowledge required.

In the later chapters of the book, a number of representative circuits (with component values) have been included together with sufficient information to allow you to adapt and modify the circuits for your own use. These circuits can be used to form the basis of your own practical investigations or they can be combined together in more complex circuits.

Finally, you can learn a great deal from building, testing and modifying simple circuits. To do this you will need access to a few basic tools and some minimal test equipment. Your first purchase should be a simple multi-range meter, either digital or analogue. This instrument will allow you to measure the voltages and currents present in your circuits and compare them with predicted values. If you are attending a formal course of instruction and have access to an electronics laboratory, do make full use of it!

A note for teachers and lecturers

The book is ideal for students following formal courses (e.g. GCSE, AS, A-level, AVCE, BTEC, City and Guilds, RSA, etc.) in schools, sixth-form colleges, and further/higher education colleges. It is equally well suited for use as a text that can support distance or flexible learning and for those who may need a 'refresher' before studying electronics at a higher level.

While the book assumes little previous knowledge students need to be able to manipulate basic formulae and understand some simple trigonometry to follow the numerical examples. A study of mathematics to GCSE level (or beyond) will normally be adequate to satisfy this requirement. However, an appendix has been added specifically to support students who may have difficulty with mathematics. Students will require a scientific calculator in order to tackle the end-of-chapter problems as well as the revision problems that appear at the end of the book.

We have also included 21 sample coursework assignments. These are open-ended and can be modified or extended to suit the requirements of the particular awarding body. The assignments have been divided into those that are broadly at Level 2 and those that are at Level 3. In order to give reasonable coverage of the subject, students should normally be expected to complete between four and five of these assignments. Teachers can differentiate students' work by mixing assignments from the two levels. In order to challenge students, minimal information should be given to students at the start of each assignment. The aim should be that of giving students 'food for thought' and encouraging them to develop their own solutions and interpretation of the topic.

Where this text is to be used to support formal teaching it is suggested that the chapters should be followed broadly in the order that they appear with the notable exception of Chapter 14. Topics from this chapter should be introduced at an early stage in order to support formal lab work. Assuming a notional delivery time of 4.5 hours per week, the material contained in this book (together with supporting laboratory exercises and assignments) will require approximately two academic terms (i.e. 24 weeks) to deliver in which the total of 90 hours of study time should be divided equally into theory (supported by problem solving) and practical (laboratory and assignment work). The recommended four or five assignments will require about 25 to 30 hours of student work to complete. Finally, when constructing a teaching programme it is, of course, essential to check that you fully comply with the requirements of the awarding body concerning assessment and that the syllabus coverage is adequate.

A word about safety

When working on electronic circuits, personal safety (both yours and of those around you) should be paramount in everything that you do. Hazards can exist within many circuits – even those that, on the face of it, may appear to be totally safe. Inadvertent misconnection of a supply, incorrect earthing, reverse connection of a high-value electrolytic capacitor, and incorrect component substitution can all result in serious hazards to personal safety as a consequence of fire, explosion or the generation of toxic fumes.

Potential hazards can be easily recognized and it is well worth making yourself familiar with them so that pitfalls can be avoided. The most important point to make is that electricity acts very quickly; you should always think carefully before taking any action where mains or high voltages (i.e. those over 50 V, or so) are concerned. Failure to observe this simple precaution may result in the very real risk of electric shock.

Voltages in many items of electronic equipment, including all items which derive their power from the a.c. mains supply, are at a level which can cause sufficient current flow in the body to disrupt normal operation of the heart. The threshold will be even lower for anyone with a defective heart. Bodily contact with mains or high-voltage circuits can thus be lethal. The most severe path for electric current within the body (i.e. the one that is most likely to stop the heart) is that which exists from one hand to the other. The hand-to-foot path is also dangerous but somewhat less danger-ous than the hand-to-hand path.

Before you start to work on an item of electronic equipment, it is essential not only to switch off but to disconnect the equipment at the mains by removing the mains plug. If you have to make measurements or carry out adjustments on a piece of working (or 'live') equipment, a useful precaution is that of using one hand only to perform the adjustment or to make the measure-ment. Your 'spare' hand should be placed safely away from contact with anything metal (including the chassis of the equipment which may, or may not, be earthed).

The severity of electric shock depends upon several factors including the magnitude of the current, whether it is alternating or direct current, and its precise path through the body. The magnitude of the current depends upon the voltage which is applied and the resistance of the body. The electrical energy developed in the body will depend upon the time for which the current flows. The duration of contact is also crucial in determining the eventual physiological effects of the shock. As a rough guide, and assuming that the voltage applied is from the 250 V 50 Hz a.c. mains supply, the following effects are typical:

Current	Physiological effect
less than 1 mA	Not usually noticeable
1 mA to 2 mA	Threshold of perception (a slight tingle may be felt)
2 mA to 4 mA	Mild shock (effects of current flow are felt)
4 mA to 10 mA	Serious shock (shock is felt as pain)
10 mA to 20 mA	Motor nerve paralysis may occur (unable to let go)
20 mA to 50 mA	Respiratory control inhibited (breathing may stop)
more than 50 mA	Ventricular fibrillation of heart muscle (heart failure)

It is important to note that the figures are quoted as a guide – there have been cases of lethal shocks resulting from contact with much lower voltages and at relatively small values of current. It is also worth noting that electric shock is often accom-panied by burns to the skin at the point of contact. These burns may be extensive and deep even when there may be little visible external damage to the skin.

The upshot of all this is simply that **any potential in excess of 50 V should be considered dangerous**. Lesser potentials may, under unusual circumstances, also be dangerous. As such, it is wise to **get into the habit of treating all electrical and electronic circuits with great care**.

1

Electrical fundamentals

This chapter has been designed to provide you with the background knowledge required to help you understand the concepts introduced in the later chapters. If you have studied electrical science or electrical principles beyond GCSE or GNVQ Intermediate Level then you will already be familiar with many of these concepts. If, on the other hand, you are returning to study or are a newcomer to electrical technology this chapter will help you get up to speed.

Fundamental units

The units that we now use to describe such things as length, mass and time are standardized within the International System of Units. This SI system is based upon the seven **fundamental units** (see Table 1.1).

Derived units

All other units are derived from these seven fundamental units. These **derived units** generally have their own names and those commonly encountered in electrical circuits are summarized in Table 1.2 together with the physical quantities to which they relate.

Table 1.1   SI units

Quantity	Unit	Abbreviation
Current	ampere	A
Length	metre	m
Luminous intensity	candela	cd
Mass	kilogram	kg
Temperature	Kelvin	K
Time	second	s
Matter	mol	mol

(Note that 0 K is equal to $-273°$C and an **interval** of 1 K is the same as an **interval** of 1°C.)

If you find the exponent notation shown in the table a little confusing, just remember that V^{-1} is simply $1/V$, s^{-1} is $1/s$, m^{-2} is $1/m^2$, and so on.

Example 1.1

The unit of flux density (the tesla) is defined as the magnetic flux per unit area. Express this in terms of the fundamental units.

Solution

The SI unit of flux is the weber (Wb). Area is directly proportional to length squared and, expressed in terms of the fundamental SI units, this is square metres (m^2). Dividing the flux (Wb) by the area (m^2) gives Wb/m^2 or $Wb\,m^{-2}$. Hence, in terms of the fundamental SI units, the tesla is expressed in $Wb\,m^{-2}$.

Example 1.2

The unit of electrical potential, the volt (V), is defined as the difference in potential between two

Table 1.2   Electrical quantities

Quantity	Derived unit	Abbreviation	Equivalent (in terms of fundamental units)
Capacitance	farad	F	$A\,s\,V^{-1}$
Charge	coulomb	C	$A\,s$
Energy	joule	J	$N\,m$
Force	newton	N	$kg\,m\,s^{-1}$
Frequency	hertz	Hz	s^{-1}
Illuminance	lux	lx	$lm\,m^{-2}$
Inductance	henry	H	$V\,s\,A^{-1}$
Luminous flux	lumen	lm	$cd\,sr$
Magnetic flux	weber	Wb	$V\,s$
Potential	volt	V	$W\,A^{-1}$
Power	watt	W	$J\,s^{-1}$
Resistance	ohm	Ω	$V\,A^{-1}$

Figure 1.1   One cycle of a sine wave

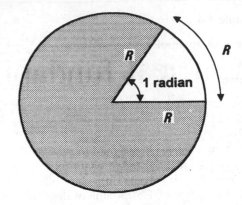

Figure 1.2   Definition of the radian

points in a conductor which, when carrying a current of one amp (A), dissipates a power of one watt (W). Express the volt (V) in terms of joules (J) and coulombs (C).

Solution

In terms of the derived units:

$$\text{volts} = \frac{\text{watts}}{\text{amperes}} = \frac{\text{joules/seconds}}{\text{amperes}}$$

$$= \frac{\text{joules}}{\text{amperes} \times \text{seconds}} = \frac{\text{joules}}{\text{coulombs}}$$

Note that: watts = joules/seconds and coulombs = amperes × seconds

In terms of units:

$$V = \frac{W}{A} = \frac{J/s}{A} = \frac{J}{As} = \frac{J}{C}$$

Hence one volt is equivalent to one joule per coulomb.

Measuring angles

You might think it strange to be concerned with angles in electrical circuits. The reason is simply that, in analogue and a.c. circuits, signals are based on repetitive waves (often sinusoidal in shape). We can refer to a point on such a wave in one of two basic ways, either in terms of the time from the start of the cycle or in terms of the angle (a cycle starts at 0° and finishes as 360° (see Fig. 1.1)). In practice, it is often more convenient to use angles rather than time, however, the two methods of measurement are interchangeable.

In electrical circuits, angles are measured in either degrees or radians (both of which are strictly dimensionless units). You will doubtless already be familiar with angular measure in degrees where one complete circular revolution is equivalent to an angular change of 360°. The alternative method of measuring angles, the **radian**, is defined somewhat differently. It is the angle subtended at the centre of a circle by an arc having length which is equal to the radius of the circle (see Fig. 1.2).

It is often necessary to convert from radians to degrees, and vice versa. A complete circular revolution is equivalent to a rotation of 360° or 2π radians (note that π is approximately equal to 3.142). Thus one radian is equivalent to $360/2\pi$ degrees (or approximately 57.3°). The following rules should assist you when it is necessary to convert angles expressed in degrees to radians and vice versa.

(a) To convert from degrees to radians, divide by 57.3.
(b) To convert from radians to degrees, multiply by 57.3.

Example 1.3

Express a quarter of a cycle revolution in terms of:

(a) degrees;
(b) radians.

Solution

(a) There are 360° in one complete cycle (i.e. one revolution). Hence there are 360/4 or 90° in one quarter of a cycle.
(b) There are 2π radians in one complete cycle (i.e. one revolution). Hence there are $2\pi/4$ or $\pi/2$ radians in one quarter of a cycle.

Example 1.4

Express an angle of 215° in radians.

Solution

To convert from degrees to radians, divide by 57.3. Hence 215° is equivalent to 215/57.3 = 3.75 radians.

Example 1.5

Express an angle of 2.5 radians in degrees.

Solution

To convert from radians to degrees, multiply by 57.3. Hence 2.5 radians is equivalent to $2.5 \times 57.3 = 143.25°$.

Electrical units and symbols

You will find that the following units and symbols are commonly encountered in electrical circuits. It is important to get to know these units and also be able to recognize their abbreviations and symbols (see Table 1.3).

Multiples and sub-multiples

Unfortunately, many of the derived units are somewhat cumbersome for everyday use but we can make life a little easier by using a standard range of multiples and sub-multiples (see Table 1.4).

Example 1.6

An indicator lamp requires a current of 0.075 A. Express this in mA.

Solution

We can express the current in mA (rather than in A) by simply moving the decimal point three places to the right. Hence 0.075 A is the same as 75 mA.

Example 1.7

A medium-wave radio transmitter operates on a frequency of 1495 kHz. Express its frequency in MHz.

Table 1.3 Electrical units

Unit	Abbrev.	Symbol	Notes
Ampere	A	I	Unit of electric current (a current of 1 A flows in a conductor when a charge of 1 C is transported in a time interval of 1 s)
Coulomb	C	Q	Unit of electric charge or quantity of electricity
Farad	F	C	Unit of capacitance (a capacitor has a capacitance of 1 F when a charge of 1 C results in a potential difference of 1 V across its plates)
Henry	H	L	Unit of inductance (an inductor has an inductance of 1 H when an applied current changing uniformly at a rate of 1 A/s produces a potential difference of 1 V across its terminals)
Hertz	Hz	f	Unit of frequency (a signal has a frequency of 1 Hz if one complete cycle occurs in a time interval of 1 s)
Joule	J	E	Unit of energy
Ohm	Ω	R	Unit of resistance
Second	s	t	Unit of time
Siemen	S	G	Unit of conductance (the reciprocal of resistance)
Tesla	T	B	Unit of magnetic flux density (a flux density of 1 T is produced when a flux of 1 Wb is present over an area of 1 square metre)
Volt	V	V	Unit of electric potential (e.m.f. or p.d.)
Watt	W	P	Unit of power (equal to 1 J of energy consumed in a time of 1 s)
Weber	Wb	Φ	Unit of magnetic flux

Table 1.4 Multiples and sub-multiples

Prefix	Abbrev.	Multiplier	
tera	T	10^{12}	(= 1 000 000 000 000)
giga	G	10^{9}	(= 1 000 000 000)
mega	M	10^{6}	(= 1 000 000)
kilo	k	10^{3}	(= 1 000)
(none)	(none)	10^{0}	(= 1)
centi	c	10^{-2}	(= 0.01)
milli	m	10^{-3}	(= 0.001)
micro	μ	10^{-6}	(= 0.000 001)
nano	n	10^{-9}	(= 0.000 000 001)
pico	p	10^{-12}	(= 0.000 000 000 001)

Solution

To express the frequency in MHz rather than kHz we need to move the decimal point three places to the left. Hence 1495 kHz is equivalent to 1.495 MHz.

Example 1.8

A capacitor has a value of 27 000 pF. Express this in μF.

Solution

To express the value in μF rather than pF we need to move the decimal point six places to the left. Hence 27 000 pF is equivalent to 0.027 μF (note that we have had to introduce an extra zero before the 2 and after the decimal point).

Exponent notation

Exponent notation (or **scientific notation**) is useful when dealing with either very small or very large quantities. It is well worth getting to grips with this notation as it will allow you to simplify quantities before using them in formulae.

Exponents are based on **powers of ten**. To express a number in exponent notation the number is split into two parts. The first part is usually a number in the range 0.1 to 100 while the second part is a multiplier expressed as a power of ten. For example, 251.7 can be expressed as 2.517×100, i.e. 2.517×10^{2}. It can also be expressed as 0.2517×1000, i.e. 0.2517×10^{3}. In both cases the exponent is the same as the number of noughts in the multiplier (i.e. 2 in the first case and 3 in the second case). To summarize:

$$251.7 = 2.517 \times 10^{2} = 0.2517 \times 10^{3}$$

As a further example, 0.01825 can be expressed as 1.825/100, i.e. 1.825×10^{-2}. It can also be expressed as 18.25/1000, i.e. 18.25×10^{-3}. Again, the exponent is the same as the number of noughts but the minus sign is used to denote a fractional multiplier. To summarize:

$$0.01825 = 1.825 \times 10^{-2} = 18.25 \times 10^{-3}$$

Example 1.9

A current of 7.25 mA flows in a circuit. Express this current in amperes using exponent notation.

Solution

$1 \, \text{mA} = 1 \times 10^{-3} \, \text{A}$ thus $7.25 \, \text{mA} = 7.25 \times 10^{-3} \, \text{A}$

Example 1.10

A voltage of $3.75 \times 10^{-6} \, \text{V}$ appears at the input of an amplifier. Express this voltage in volts using exponent notation.

Solution

$1 \times 10^{-6} \, \text{V} = 1 \, \mu\text{V}$ thus $3.75 \times 10^{-6} \, \text{V} = 3.75 \, \mu\text{V}$

Multiplication and division using exponents

Exponent notation really comes into its own when values have to be multiplied or divided. When multiplying two values expressed using exponents, it is simply necessary to add the exponents. As an example:

$$(2 \times 10^{2}) \times (3 \times 10^{6}) = (2 \times 3) \times 10^{(2+6)} = 6 \times 10^{8}$$

Similarly, when dividing two values which are expressed using exponents, it is simply necessary to subtract the exponents. As an example:

$$(4 \times 10^{6}) \div (2 \times 10^{4}) = 4/2 \times 10^{(6-4)} = 2 \times 10^{2}$$

In either case it is essential to take care to express the units, multiples and sub-multiples in which you are working.

Example 1.11

A current of 3 mA flows in a resistance of 33 kΩ. Determine the voltage dropped across the resistor.

Solution

Voltage is equal to current multiplied by resistance (see page 6). Thus:

$V = I \times R = 3\,\text{mA} \times 33\,\text{k}\Omega$

Expressing this using exponent notation gives:

$V = (3 \times 10^{-3}) \times (33 \times 10^{3})$ volts

Separating the exponents gives:

$V = 3 \times 33 \times 10^{-3} \times 10^{3}$ volts

Thus $V = 99 \times 10^{(-3+3)} = 99 \times 10^{0} = 99 \times 1 = 99$ V

Example 1.12

A current of 45 µA flows in a circuit. What charge is transferred in a time interval of 20 ms?

Solution

Charge is equal to current multiplied by time (see the definition of the ampere on page 3). Thus:

$Q = it = 45\,\mu\text{A} \times 20\,\text{ms}$

Expressing this in exponent notation gives:

$Q = (45 \times 10^{-6}) \times (20 \times 10^{-3})$ coulomb

Separating the exponents gives:

$Q = 45 \times 20 \times 10^{-6} \times 10^{-3}$ coulomb

Thus $Q = 900 \times 10^{(-6-3)} = 900 \times 10^{(-9)} = 900\,\text{nC}$

Example 1.13

A power of 300 mW is dissipated in a circuit when a voltage of 1500 V is applied. Determine the current supplied to the circuit.

Solution

Current is equal to power divided by voltage (see page 8). Thus:

$I = P/V = 300\,\text{mW}/1500\,\text{V}$ amperes

Expressing this in exponent notation gives:

$I = (300 \times 10^{-3})/(1.5 \times 10^{3})$ amperes

Separating the exponents gives:

$I = 300/1.5 \times 10^{-3}/10^{3}$ amperes

$I = 300/1.5 \times 10^{-3} \times 10^{-3}$ amperes

Thus $I = 200 \times 10^{(-3-3)} = 200 \times 10^{(-6)} = 200\,\mu\text{A}$

Conductors and insulators

Electric current is the name given to the flow of **electrons** (or negative charge carriers). Electrons orbit around the nucleus of atoms just as the earth orbits around the sun (see Fig. 1.3). Electrons are held in one or more **shells**, constrained to their orbital paths by virtue of a force of attraction towards the nucleus which contains an equal number of **protons** (positive charge carriers). Since like charges repel and unlike charges attract, negatively charged electrons are attracted to the positively charged nucleus. A similar principle can be demonstrated by observing the attraction between two permanent magnets; the two north poles of the magnets will repel each other, while a north and south pole will attract. In the same way, the unlike charges of the negative electron and the positive proton experience a force of mutual attraction.

The outer shell electrons of a **conductor** can be reasonably easily interchanged between adjacent atoms within the **lattice** of atoms of which the substance is composed. This makes it possible for the material to conduct electricity. Typical examples of conductors are metals such as copper, silver, iron and aluminium. By contrast, the outer shell electrons of an **insulator** are firmly bound to their parent atoms and virtually no interchange of electrons is possible. Typical examples of insulators are plastics, rubber and ceramic materials.

Figure 1.3 A single atom of helium (He) showing its two electrons in orbit around its nucleus

Figure 1.4   Simple circuit to illustrate the relationship between voltage (V), current (I) and resistance (R) (note the direction of conventional current flow from positive to negative)

Voltage and resistance

The ability of an energy source (e.g. a battery) to produce a current within a conductor may be expressed in terms of **electromotive force** (e.m.f.). Whenever an e.m.f. is applied to a circuit **a potential difference** (p.d.) exists. Both e.m.f. and p.d. are measured in volts (V). In many practical circuits there is only one e.m.f. present (the battery or supply) whereas a p.d. will be developed across each component present in the circuit.

The conventional flow of current in a circuit is from the point of more positive potential to the point of greatest negative potential (note that electrons move in the *opposite* direction!). **Direct current** results from the application of a direct e.m.f. (derived from batteries or d.c. supply rails). An essential characteristic of such supplies is that the applied e.m.f. does not change its polarity (even though its value might be subject to some fluctuation).

For any conductor, the current flowing is directly proportional to the e.m.f. applied. The current flowing will also be dependent on the physical dimensions (length and cross-sectional area) and material of which the conductor is composed. The amount of current that will flow in a conductor when a given e.m.f. is applied is inversely proportional to its **resistance**. Resistance, therefore, may be thought of as an opposition to current flow; the higher the resistance the lower the current that will flow (assuming that the applied e.m.f. remains constant).

Ohm's law

Provided that temperature does not vary, the ratio of p.d. across the ends of a conductor to the current flowing in the conductor is a constant. This relationship is known as Ohm's law and it leads to the relationship:

$$V/I = \text{a constant} = R$$

where V is the potential difference (or voltage drop) in volts (V), I is the current in amperes (A), and R is the resistance in ohms (Ω) (see Fig. 1.4).

The formula may be arranged to make V, I or R the subject, as follows:

$$V = I \times R \qquad I = V/R \qquad \text{and} \qquad R = V/I$$

The triangle shown in Fig. 1.5 should help you remember these three important relationships. It is important to note that, when performing calculations of currents, voltages and resistances in practical circuits it is seldom necessary to work

Figure 1.5   Triangle showing the relationship between V, I and R

with an accuracy of better than ±1% simply because component tolerances are invariably somewhat greater than this. Furthermore, in calculations involving Ohm's law, it is sometimes convenient to work in units of kΩ and mA (or MΩ and μA) in which case potential differences will be expressed directly in V.

Example 1.14

A 12 Ω resistor is connected to a 6 V battery. What current will flow in the resistor?

Solution

Here we must use $I = V/R$ (where $V = 6$ V and $R = 12$ Ω):

$I = V/R = 6\,\text{V}/12\,\Omega = 0.5\,\text{A}$ (or 500 mA)

Hence a current of 500 mA will flow in the resistor.

Example 1.15

A current of 100 mA flows in a 56 Ω resistor. What voltage drop (potential difference) will be developed across the resistor?

Solution

Here we must use $V = I \times R$ and ensure that we work in units of volts (V), amperes (A) and ohms (Ω).

$V = I \times R = 0.1\,\text{A} \times 56\,\Omega = 5.6\,\text{V}$

(Note that 100 mA is the same as 0.1 A.)

Hence a p.d. of 5.6 V will be developed across the resistor.

Example 1.16

A voltage drop of 15 V appears across a resistor in which a current of 1 mA flows. What is the value of the resistance?

Solution

$R = V/I = 15\,\text{V}/0.001\,\text{A} = 15\,000\,\Omega = 15\,\text{k}\Omega$

Note that it is often more convenient to work in units of mA and V which will produce an answer directly in kΩ, i.e.

$R = V/I = 15\,\text{V}/1\,\text{mA} = 15\,\text{k}\Omega$

Resistance and resistivity

The resistance of a metallic conductor is directly proportional to its length and inversely proportional to its area. The resistance is also directly proportional to its **resistivity** (or **specific resistance**). Resistivity is defined as the resistance measured between the opposite faces of a cube having sides of 1 cm.

The resistance, R, of a conductor is thus given by the formula:

$R = \rho \times l/A$

where R is the resistance (Ω), ρ is the resistivity (Ω m), l is the length (m), and A is the area (m²).

Table 1.5 shows the electrical properties of various metals.

Example 1.17

A coil consists of an 8 m length of annealed copper wire having a cross-sectional area of 1 mm². Determine the resistance of the coil.

Solution

We will use the formula, $R = \rho l/A$.

The value of ρ for copper is 1.724×10^{-8} Ω m given in Table 1.5 which shows the properties of common metallic conductors. The length of the wire is 4 m while the area is 1 mm² or 1×10^{-6} m² (note

Table 1.5 Properties of common metals

Metal	Resistivity (at 20°C) (Ωm)	Relative conductivity (copper = 1)	Temperature coefficient of resistance (per °C)
Silver	1.626×10^{-8}	1.06	0.0041
Copper (annealed)	1.724×10^{-8}	1.00	0.0039
Copper (hard drawn)	1.777×10^{-8}	0.97	0.0039
Aluminium	2.803×10^{-8}	0.61	0.0040
Mild steel	1.38×10^{-7}	0.12	0.0045
Lead	2.14×10^{-7}	0.08	0.0040

that it is important to be consistent in using units of metres for length and square metres for area).

Hence the resistance of the coil will be given by:

$$R = \frac{1.724 \times 10^{-8} \times 8}{1 \times 10^{-6}}$$

Thus $R = 13.792 \times 10^{-2}$ or $0.137\,92\,\Omega$.

Example 1.18

A wire having a resistivity of $1.6 \times 10^{-8}\,\Omega\,m$, length $20\,m$ and cross-sectional area $1\,mm^2$ carries a current of $5\,A$. Determine the voltage drop between the ends of the wire.

Solution

First we must find the resistance of the wire (as in Example 1.17):

$$R = \rho l/A = \frac{1.6 \times 10^{-8} \times 20}{1 \times 10^{-6}} = 0.32\,\Omega$$

The voltage drop can now be calculated using Ohm's law:

$$V = I \times R = 5\,A \times 0.32\,\Omega = 1.6\,V$$

Hence a potential of 1.6 V will be dropped between the ends of the wire.

Energy and power

At first you may be a little confused about the difference between energy and power. Energy is the ability to do work while power is the rate at which work is done. In electrical circuits, energy is supplied by batteries or generators. It may also be stored in components such as capacitors and inductors. Electrical energy is converted into various other forms of energy by components such as resistors (producing heat), loudspeakers (producing sound energy) and light emitting diodes (producing light).

The unit of energy is the joule (J). Power is the rate of use of energy and it is measured in watts (W). A power of 1 W results from energy being used at the rate of 1 J per second. Thus:

$$P = E/t$$

where P is the power in watts (W), E is the energy in joules (J), and t is the time in seconds (s).

The power in a circuit is equivalent to the product of voltage and current. Hence:

Figure 1.6   Triangle showing the relationship between P, I and V

$$P = I \times V$$

where P is the power in watts (W), I is the current in amperes (A), and V is the voltage in volts (V).

The formula may be arranged to make P, I or V the subject, as follows:

$$P = I \times V \qquad I = P/V \qquad \text{and} \qquad V = P/I$$

The triangle shown in Fig. 1.6 should help readers remember these relationships.

The relationship, $P = I \times V$, may be combined with that which results from Ohm's law ($V = I \times R$) to produce two further relationships. First, substituting for V gives:

$$P = I \times (I \times R) = I^2 R$$

Secondly, substituting for I gives:

$$P = (V/R) \times V = V^2/R$$

Example 1.19

A current of 1.5 A is drawn from a 3 V battery. What power is supplied? Here we must use $P = I \times V$ (where $I = 1.5\,A$ and $V = 3\,V$):

Solution

$$P = I \times V = 1.5\,A \times 3\,V = 4.5\,W$$

Hence a power of 4.5 W is supplied.

Example 1.20

A voltage drop of 4 V appears across a resistor of $100\,\Omega$. What power is dissipated in the resistor?

Solution

Here we use $P = V^2/R$ (where $V = 4\,V$ and $R = 100\,\Omega$):

$P = V^2/R = (4\,\text{V} \times 4\,\text{V})/100\,\Omega = 0.16\,\text{W}$

Hence the resistor dissipates a power of 0.16 W (or 160 mW).

Example 1.21

A current of 20 mA flows in a 1 k Ω resistor. What power is dissipated in the resistor?

Solution

Here we use $P = I^2 \times R$ but, to make life a little easier, we will work in mA and kΩ (in which case the answer will be in mW):

$P = I^2 \times R = (20\,\text{mA} \times 20\,\text{mA}) \times 1\,\text{k}\Omega$

$\quad = 400\,\text{mW}$

Thus a power of 400 mW is dissipated in the resistor.

Electrostatics

If a conductor has a deficit of electrons, it will exhibit a net positive charge. If, on the other hand, it has a surplus of electrons, it will exhibit a net positive charge. An imbalance in charge can be produced by friction (removing or depositing electrons using materials such as silk and fur, respectively) or induction (by attracting or repelling electrons using a second body which is, respectively, positively or negatively charged).

Force between charges

Coulomb's law states that, if charged bodies exist at two points, the force of attraction (if the charges are of opposite charge) or repulsion (if of like charge) will be proportional to the product of the magnitude of the charges divided by the square of their distance apart. Thus:

$$F = \frac{kQ_1 Q_2}{r^2}$$

where Q_1 and Q_2 are the charges present at the two points (in coulombs), r the distance separating the two points (in metres), F is the force (in newtons), and k is a constant depending upon the medium in which the charges exist.

In vacuum or 'free space',

$$k = \frac{1}{4\pi\epsilon_0}$$

where ϵ_0 is the **permittivity of free space** (8.854$\times$ 10^{-12} C/N m^2).

Combining the two previous equations gives:

$$F = \frac{Q_1 Q_2}{4\pi\epsilon_0 r^2}$$

or

$$F = \frac{Q_1 Q_2}{4\pi 8.854 \times 10^{-12} r^2} \quad \text{newtons}$$

Electric fields

The force exerted on a charged particle is a manifestation of the existence of an electric field. The electric field defines the direction and magnitude of a force on a charged object. The field itself is invisible to the human eye but can be drawn by constructing lines which indicate the motion of a free positive charge within the field; the number of field lines in a particular region being used to indicate the relative strength of the field at the point in question.

Figures 1.7 and 1.8 show the electric fields between unlike and like charges while Fig. 1.9 shows the field which exists between two charged parallel plates (note the 'fringing' which occurs at the edges of the plates).

Electric field strength

The strength of an electric field (E) is proportional to the applied potential difference and inversely

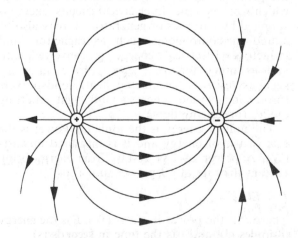

Figure 1.7   Electric fields between unlike electric charges

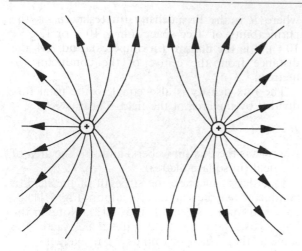

Figure 1.8   Electric fields between like electric charges

Figure 1.9   Electric field between two charged parallel plates

proportional to the distance between the two conductors. The electric field strength is given by:

$$E = V/d$$

where E is the electric field strength (V/m), V is the applied potential difference (V) and d is the distance (m).

Example 1.22

Two parallel conductors are separated by a distance of 25 mm. Determine the electric field strength if they are fed from a 600 V d.c. supply.

Solution

The electric field strength will be given by:

$$E = V/d = 600/25 \times 10^{-3} = 24 \text{ kV/m}$$

Electromagnetism

When a current flows through a conductor a magnetic field is produced in the vicinity of the conductor. The magnetic field is invisible but its presence can be detected using a compass needle (which will deflect from its normal north–south position). If two current-carrying conductors are placed in the vicinity of one another, the fields will interact with one another and the conductors will experience a force of attraction or repulsion (depending upon the relative direction of the two currents).

Force between current-carrying conductors

The mutual force which exists between two parallel current-carrying conductors will be proportional to the product of the currents in the two conductors and the length of the conductors but inversely proportional to their separation. Thus:

$$F = \frac{kI_1I_2l}{d}$$

where I_1 and I_2 are the currents in the two conductors (in amps), l is the parallel length of the conductors (in metres), d is the distance separating the two conductors (in metres), F is the force (in newtons), and k is a constant depending upon the medium in which the charges exist.

In vacuum or 'free space',

$$k = \frac{\mu_0}{2\pi}$$

where μ_0 is a constant known as the **permeability of free space** (12.57×10^{-7} H/m).

Combining the two previous equations gives:

$$F = \frac{\mu_0 I_1 I_2 l}{2\pi d}$$

or

$$F = \frac{4\pi \times 10^{-7} I_1 I_2 l}{2\pi d} \text{ newtons}$$

or

$$F = \frac{2 \times 10^{-7} I_1 I_2 l}{d} \text{ newtons}$$

Magnetic fields

The field surrounding a straight current-carrying conductor is shown in Fig. 1.10. The magnetic field defines the direction of motion of a free north pole within the field. In the case of Fig. 1.10, the lines of flux are concentric and the direction of the field (determined by the direction of current flow) is given by the right-hand screw rule.

Magnetic field strength

The strength of a magnetic field is a measure of the density of the flux at any particular point. In the case of Fig. 1.10, the field strength will be proportional to the applied current and inversely proportional to the perpendicular distance from the conductor. Thus

$$B = \frac{kI}{d}$$

where B is the magnetic flux density (in tesla), I is the current (in amperes), d is the distance from the conductor (in metres), and k is a constant. Assuming that the medium is vacuum or 'free space', the density of the magnetic flux will be given by:

$$B = \frac{\mu_0 I}{2\pi d}$$

where B is the flux density (in tesla), μ_0 is the permeability of 'free space' ($4\pi \times 10^{-7}$ or 12.57×10^{-7}), I is the current (in amperes), and d is the distance from the centre of the conductor (in metres).

The flux density is also equal to the total flux divided by the area of the field. Thus:

$$B = \Phi/A$$

where Φ is the flux (in webers) and A is the area of the field (in square metres).

In order to increase the strength of the field, a conductor may be shaped into a loop (Fig. 1.11) or coiled to form a solenoid (Fig. 1.12). Note, in the latter case, how the field pattern is exactly the same as that which surrounds a bar magnet.

Example 1.23

Determine the flux density produced at a distance of 50 mm from a straight wire carrying a current of 20 A.

Solution

Applying the formula $B = \mu_0 I/2\pi d$ gives:

$$B = \frac{12.57 \times 10^{-7} \times 20}{2 \times 3.142 \times 50 \times 10^{-3}} = \frac{251.40 \times 10^{-7}}{314.20 \times 10^{-3}}$$

$$= 0.8 \times 10^{-4} \text{ tesla}$$

thus $B = 800 \times 10^{-6}$ T or $B = 80\,\mu\text{T}$.

Example 1.24

A flux density of 2.5 mT is developed in free space over an area of 20 cm^2. Determine the total flux.

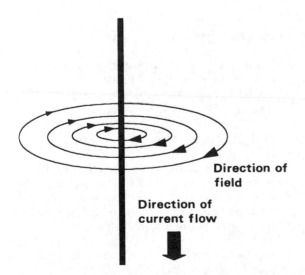

Figure 1.10   Magnetic field surrounding a straight current-carrying conductor

Direction of field

Direction of current flow

Current, I Current, I

Figure 1.11   Forming a conductor into a loop increases the strength of the magnetic field in the centre of the loop

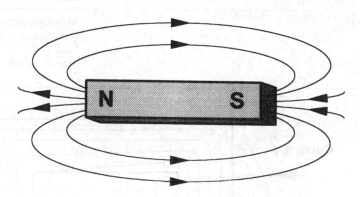

Figure 1.12   Magnetic field around a solenoid

Figure 1.13   An electric circuit

Solution

Re-arranging the formula $B = \Phi/A$ to make Φ the subject gives $\Phi = BA$ thus $\Phi = (2.5 \times 10^{-3}) \times (20 \times 10^{-4}) = 50 \times 10^{-7}$ Wb $= 5\,\mu$Wb.

Magnetic circuits

Materials such as iron and steel possess considerably enhanced magnetic properties. Hence they are employed in applications where it is necessary to increase the flux density produced by an electric current. In effect, magnetic materials allow us to channel the electric flux into a 'magnetic circuit', as shown in Fig. 1.14.

In the circuit of Fig. 1.14 the **reluctance** of the magnetic core is analogous to the resistance present in the electric circuit shown in Fig. 1.13. We can make the following comparisons between the two types of circuit (see Table 1.6):

In practice, not all of the magnetic flux produced in a magnetic circuit will be concentrated within the core and some 'leakage flux' will appear in the surrounding free space (as shown in Fig. 1.15). Similarly, if a gap appears within the magnetic

Figure 1.14 A magnetic circuit

Table 1.6 Comparison of electric and magnetic circuits

Electric circuit (Fig. 1.13)	*Magnetic circuit* (Fig. 1.14)
Electromotive force, e.m.f. $= V$	Magnetomotive force, m.m.f. $= N \times I$
Resistance $= R$	Reluctance $= S$
Current $= I$	Flux $= \Phi$
e.m.f. $=$ current $\times$ resistance	m.m.f. $=$ flux $\times$ reluctance
$V = IR$	$NI = S\Phi$

Figure 1.15 Leakage flux in a magnetic circuit

circuit, the flux will tend to spread out as shown in Fig. 1.16. This effect is known as 'fringing'.

Reluctance and permeability

The reluctance of a magnetic path is directly proportional to its length and inversely proportional to its area. The reluctance is also inversely proportional to the **absolute permeability** of the magnetic material. Thus

$$S = \frac{l}{\mu A}$$

where S is the reluctance of the magnetic path, l is the length of the path (in metres), A is the cross-sectional area of the path (in square metres), and μ is the absolute permeability of the magnetic material.

Figure 1.16 Fringing of the magnetic flux in an air gap

Now the absolute permeability of a magnetic material is the product of the permeability of free space (μ_0) and the **relative permeability** of the magnetic medium (μ_r). Thus

$$\mu = \mu_0 \times \mu_r \text{ and } S = \frac{l}{\mu_0 \mu_r A}$$

The permeability of a magnetic medium is a measure of its ability to support magnetic flux and it is equal to the ratio of flux density (B) to **magnetizing force (H)**. Thus

$$\mu = \frac{B}{H}$$

where B is the flux density (in tesla) and H is the magnetizing force (in ampere/metre).

The magnetizing force (H) is proportional to the product of the number of turns and current but inversely proportional to the length of the magnetic path. Thus:

$$H = \frac{N \times I}{l}$$

where H is the magnetizing force (in ampere/metre), N is the number of turns, I is the current (in amperes), and l is the length of the magnetic path (in metres).

— — — 4.5% silicon iron
· · · · · · mild steel
— · — · cast steel
———— cast iron

Figure 1.17 *B–H* curves for four magnetic materials

B–H curves

Figure 1.17 shows four typical B–H (flux density plotted against permeability) curves for some common magnetic materials. It should be noted that each of these curves eventually flattens off due to magnetic saturation and that the slope of the curve (indicating the value of μ corresponding to a particular value of H) falls as the magnetizing force increases. This is important since it dictates the acceptable working range for a particular magnetic material when used in a magnetic circuit.

Example 1.25

Estimate the relative permeability of cast steel (see Fig. 1.18) at (a) a flux density of 0.6 T and (b) a flux density of 1.6 T.

Solution

From Fig. 1.18, the slope of the graph at any point gives the value of μ at that point. The slope can be found by constructing a tangent at the point in question and finding the ratio of vertical change to horizontal change.

(a) The slope of the graph at 0.6 T is $0.3/500 = 0.6 \times 10^{-3}$
Since $\mu = \mu_0 \mu_r$, $\mu_r = \mu/\mu_0 = 0.6 \times 10^{-3}/12.57 \times 10^{-7}$,
thus $\mu_r = 477$.

(b) The slope of the graph at 1.6 T is $0.09/1500 = 0.06 \times 10^{-3}$
Since $\mu = \mu_0 \mu_r$, $\mu_r = \mu/\mu_0 = 0.06 \times 10^{-3}/12.57 \times 10^{-7}$,
thus $\mu_r = 47.7$.

NB: This example clearly shows the effect of saturation on the permeability of a magnetic material!

Example 1.26

A coil of 800 turns is wound on a closed mild steel core having a length 600 mm and cross-sectional area 500 mm². Determine the current required to establish a flux of 0.8 mWb in the core.

Figure 1.18 B–H curve for a sample of cast steel

Solution

$B = \Phi/A = 0.8 \times 10^{-3}/500 \times 10^{-6} = 1.6\,\text{T}$

From Fig. 1.17, a flux density of 1.6 T will occur in mild steel when $H = 3500\,\text{A/m}$. The current can now be determined by re-arranging $H = (N \times I)/l$

$$I = \frac{H \times l}{N} = \frac{3500 \times 0.6}{800} = 2.625\,\text{A}$$

Circuit symbols introduced in this chapter

Fixed resistor   Single-cell battery   Multi-cell battery

Junction   No connection   Earth   Chassis ground

Figure 1.19

Important formulae introduced in this chapter

Voltage, current and resistance (Ohm's law): (page 6)

$V = IR$

Resistance and resistivity: (page 7)

$R = \rho l/A$

Charge, current and time: (page 5)

$Q = It$

Power, current and voltage: (page 8)

$P = IV$

Power, voltage and resistance: (page 8)

$P = V^2/R$

Power, current and resistance: (page 8)

$P = I^2 R$

Reluctance and permeability: (page 13)

$$S = \frac{l}{\mu A}$$

Flux and flux density:
(page 11)

$$B = \Phi/A$$

Current and magnetic field intensity:
(page 14)

$$H = \frac{NI}{l}$$

Flux, current and reluctance:
(page 13)

$$NI = S\Phi$$

Problems

1.1 Which of the following are not fundamental units; amperes, metres, coulombs, joules, hertz, kilogram?

1.2 A commonly used unit of consumer energy is the kilowatt hour (kWh). Express this in joules (J).

1.3 Express an angle of 30° in radians.

1.4 Express an angle of 0.2 radians in degrees.

1.5 A resistor has a value of 39 570 Ω. Express this in kilohms (kΩ).

1.6 An inductor has a value of 680 mH. Express this in henries (H).

1.7 A capacitor has a value of 0.002 45 μF. Express this in nanofarads (nF).

1.8 A current of 190 μA is applied to a circuit. Express this in milliamperes (mA).

1.9 A signal of 0.475 mV appears at the input of an amplifier. Express this in volts using exponent notation.

1.10 A cable has an insulation resistance of 16.5 MΩ. Express this resistance in ohms using exponent notation.

1.11 Perform the following arithmetic using exponents:

(a) $(1.2 \times 10^3) \times (4 \times 10^3)$
(b) $(3.6 \times 10^6) \times (2 \times 10^{-3})$
(c) $(4.8 \times 10^9) \div (1.2 \times 10^6)$
(d) $(9.9 \times 10^{-6}) \div (19.8 \times 10^{-3})$

1.12 Which one of the following metals is the best conductor of electricity: aluminium, copper, silver, or mild steel? Why?

1.13 A resistor of 270 Ω is connected across a 9 V d.c. supply. What current will flow?

1.14 A current of 56 μA flows in a 120 kΩ resistor. What voltage drop will appear across the resistor?

1.15 A voltage drop of 13.2 V appears across a resistor when a current of 4 mA flows in it. What is the value of the resistor?

1.16 A power supply is rated at 15 V, 1 A. What value of load resistor would be required to test the power supply at its full rated output?

1.17 A wirewound resistor is made from a 4 m length of aluminium wire ($\rho = 2.18 \times 10^{-8}$ Ω m) having a cross-sectional area of 0.2 mm². Determine the value of resistance.

1.18 A current of 25 mA flows in a 47 Ω resistor. What power is dissipated in the resistor?

1.19 A 9 V battery supplies a circuit with a current of 75 mA. What power is consumed by the circuit?

1.20 A resistor of 150 Ω is rated at 0.5 W. What is the maximum current that can be applied to the resistor without exceeding its rating?

1.21 Determine the electric field strength that appears in the space between two parallel plates separated by an air gap of 4 mm if a potential of 2.5 kV exists between them.

1.22 Determine the current that must be applied to a straight wire conductor in order to produce a flux density of 200 μT at a distance of 12 mm in free space.

1.23 A flux density of 1.2 mT is developed in free space over an area of 50 cm². Determine the total flux present.

1.24 A ferrite rod has a length of 250 mm and a diameter of 10 mm. Determine the reluctance if the rod has a relative permeability of 2500.

1.25 A coil of 400 turns is wound on a closed mild steel core having a length 400 mm and cross-sectional area 480 mm². Determine the current required to establish a flux of 0.6 mWb in the core.

(The answers to these problems appear on page 260.)

2

Passive components

This chapter introduces several of the most common types of electronic component, including resistors, capacitors and inductors. These are often referred to as **passive components** as they cannot, by themselves, generate voltage or current. An understanding of the characteristics and application of passive components is an essential prerequisite to understanding the operation of the circuits used in amplifiers, oscillators, filters and power supplies.

Resistors

The notion of resistance as opposition to current was discussed in the previous chapter. Conventional forms of resistor obey a straight line law when voltage is plotted against current (see Fig. 2.1) and this allows us to use resistors as a means of converting current into a corresponding voltage drop, and vice versa (note that doubling the applied current will produce double the voltage drop, and so on). Therefore resistors provide us with a means of controlling the currents and voltages present in electronic circuits. They can also act as **loads** to simulate the presence of a circuit during testing (e.g. a suitably rated resistor

can be used to replace a loudspeaker when an audio amplifier is being tested).

The specifications for a resistor usually include the value of resistance (expressed in ohms (Ω), kilohms (kΩ) or megohms (MΩ)), the accuracy or tolerance (quoted as the maximum permissible percentage deviation from the marked value), and the power rating (which must be equal to, or greater than, the maximum expected power dissipation).

Other practical considerations when selecting resistors for use in a particular application include temperature coefficient, noise performance, stability and ambient temperature range. Table 2.1 summarizes the properties of five of the most common types of resistor. Figures 2.2 and 2.3 show the construction of typical carbon rod (now obsolete) and carbon film resistors.

Preferred values

The value marked on the body of a resistor is not its *exact* resistance. Some minor variation in resistance value is inevitable due to production tolerance. For example, a resistor marked $100\,\Omega$ and produced within a tolerance of $\pm10\%$ will have a value which falls within the range $90\,\Omega$ to $110\,\Omega$. If a particular circuit requires a resistance of, for example, $105\,\Omega$, a $\pm10\%$ tolerance resistor of $100\,\Omega$ will be perfectly adequate. If, however, we need a component with a value of $101\,\Omega$, then it would be necessary to obtain a $100\,\Omega$ resistor with a tolerance of $\pm1\%$.

Resistors are available in several series of fixed decade values, the number of values provided with each series being governed by the tolerance involved. In order to cover the full range of resistance values using resistors having a $\pm20\%$ tolerance it will be necessary to provide six basic values (known as the **E6 series**). More values will be required in the series which offers a tolerance of $\pm10\%$ and consequently the **E12 series** provides twelve basic values. The **E24 series** for resistors of

Figure 2.1 Voltage plotted against current for two different values of resistor (note that the slope of the graph is proportional to the value of resistance)

Table 2.1 Characteristics of common types of resistor

Parameter	Resistor type				
	Carbon film	Metal film	Metal oxide	Ceramic wirewound	Vitreous wirewound
Resistance range (Ω)	10 to 10 M	1 to 1 M	10 to 1 M	0.47 to 22 k	0.1 to 22 k
Typical tolerance (%)	±5	±1	±2	±5	±5
Power rating (W)	0.25 to 2	0.125 to 0.5	0.25 to 0.5	4 to 17	2 to 4
Temperature coefficient (ppm/°C)	−250	+50 to +100	+250	+250	+75
Stability	Fair	Excellent	Excellent	Good	Good
Noise performance	Fair	Excellent	Excellent	n/a	n/a
Ambient temperature range (°C)	−45 to +125	−55 to +125	−55 to +155	−55 to +200	−55 to +200
Typical applications	General purpose	Amplifiers, test equipment, etc. requiring low-noise high-tolerance components		Power supplies, loads, high-power circuits	

Figure 2.2 Construction of a carbon rod resistor

Figure 2.3 Construction of a carbon film resistor

±5% tolerance provides no fewer than 24 basic values and, as with the E6 and E12 series, decade multiples (i.e. ×1, ×10, ×100, ×1k, ×10k, ×100k and ×1M) of the basic series. Figure 2.4 shows the relationship between the E6, E12 and E24 series.

Power ratings

Resistor power ratings are related to operating temperatures and resistors should be derated at

Figure 2.4 The E6, E12 and E24 series

high temperatures. Where reliability is important resistors should be operated at well below their nominal maximum power dissipation.

Example 2.1

A resistor has a marked value of 220 Ω. Determine the tolerance of the resistor if it has a measured value of 207 Ω.

Solution

The difference between the marked and measured values of resistance (the error) is (220 Ω − 207 Ω) = 13 Ω. The tolerance is given by:

$$\text{Tolerance} = \frac{\text{error}}{\text{marked value}} \times 100\%$$

The tolerance is thus 13/220 × 100 = 5.9%.

Example 2.2

A 9 V power supply is to be tested with a 39 Ω load resistor. If the resistor has a tolerance of 10% determine:

(a) the nominal current taken from the supply;
(b) the maximum and minimum values of supply current at either end of the tolerance range for the resistor.

Solution

(a) If a resistor of **exactly** 39 Ω is used the current will be:

$$I = V/R = 9V/39\,\Omega = 231\,\text{mA}$$

(b) The lowest value of resistance would be (39 Ω − 3.9 Ω) = 35.1 Ω. In which case the current would be:

$$I = V/R = 9V/35.1\,\Omega = 256.4\,\text{mA}$$

At the other extreme, the highest value of resistance would be (39 Ω + 3.9 Ω) = 42.9 Ω. In this case the current would be:

$$I = V/R = 9V/42.9\,\Omega = 209.8\,\text{mA}$$

Example 2.3

A current of 100 mA (±20%) is to be drawn from a 28 V d.c. supply. What value and type of resistor should be used in this application?

Solution

The value of resistance required must first be calculated using Ohm's law:

$$R = V/I = 28\,\text{V}/100\,\text{mA} = 280\,\Omega$$

The nearest preferred value from the E12 series is 270 Ω (which will actually produce a current of 103.7 mA (i.e. within ±4% of the desired value). If a resistor of ±10% tolerance is used, current will be within the range 94 mA to 115 mA (well within the ±20% accuracy specified). The power dissipated in the resistor (calculated using $P = I \times V$) will be 2.9 W and thus a component rated at 3 W (or more) will be required. This would normally be a vitreous enamel coated wirewound resistor (see Table 2.1).

Resistor markings

Carbon and metal oxide resistors are normally marked with colour codes which indicate their value and tolerance. Two methods of colour coding are in common use; one involves four coloured bands (see Fig. 2.5) while the other uses five colour bands (see Fig. 2.6).

Example 2.4

A resistor is marked with the following coloured stripes: brown, black, red, silver. What is its value and tolerance?

Solution

See Fig. 2.7.

Example 2.5

A resistor is marked with the following coloured stripes: red, violet, orange, gold. What is its value and tolerance?

Solution

See Fig. 2.8.

Example 2.6

A resistor is marked with the following coloured stripes: green, blue, black, gold. What is its value and tolerance?

Solution

See Fig. 2.9.

Example 2.7

A resistor is marked with the following coloured stripes: red, green, black, black, brown. What is its value and tolerance?

Solution

See Fig. 2.10.

1st and 2nd colour bands		3rd coloured band		4th coloured band (tolerance)	
Black	0	Silver	Multiply by 0.01	Red	± 2%
Brown	1	Gold	0.1	Gold	± 5%
Red	2	Black	1	Silver	± 10%
Orange	3	Brown	10	None	± 20%
Yellow	4	Red	100		
Green	5	Orange	1000		
Blue	6	Yellow	10000		
Violet	7	Green	100000		
Grey	8	Blue	1000000		
White	9				

Figure 2.5 Four band resistor colour code

1st, 2nd, 3rd colour bands	
Black	0
Brown	1
Red	2
Orange	3
Yellow	4
Green	5
Blue	6
Violet	7
Grey	8
White	9

4th coloured band	Multiply by
Silver	0.01
Gold	0.1
Black	1
Brown	10
Red	100
Orange	1000
Yellow	10000
Green	100000
Blue	1000000

5th coloured band (tolerance)	
Red	± 2%
Gold	± 5%
Silver	± 10%
None	± 20%

Figure 2.6   Five band resistor colour code

Brown, black, red, silver

1st band Brown = 1	2nd band Black = 0	3rd band Red = x100	4th band Silver = ± 10%

1	0	0 0	± 10%

1k ohm ± 10%

Figure 2.7

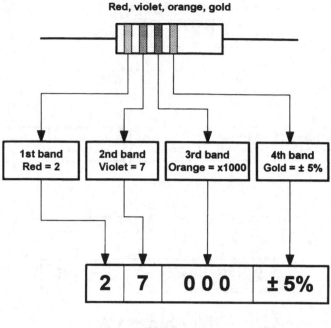

Red, violet, orange, gold

| 1st band
Red = 2 | 2nd band
Violet = 7 | 3rd band
Orange = x1000 | 4th band
Gold = ± 5% |

| 2 | 7 | 0 0 0 | ± 5% |

27k ohm ± 5%

Figure 2.8

Green, blue, black, gold

| 1st band
Green = 5 | 2nd band
Blue = 6 | 3rd band
Black = x1 | 4th band
Gold = ± 5% |

| 5 | 6 | - | ± 5% |

56 ohm ± 5%

Figure 2.9

Red, green, black, black, brown

| 1st band Red = 2 | 2nd band Green = 5 | 3rd band Black = 0 | 4th band Black = x1 | 5th band Brown = ± 1% |

| 2 | 5 | 0 | - | ± 1% |

250 ohm ± 1%

Figure 2.10

BS 1852 coding

Some types of resistor have markings based on a system of coding defined in BS 1852. This system involves marking the position of the decimal point with a letter to indicate the multiplier concerned as shown in Table 2.2. A further letter is then appended to indicate the tolerance as shown in Table 2.3.

Table 2.2   Resistor multiplier markings

Letter	Multiplier
R	1
K	1 000
M	1 000 000

Table 2.3   Resistor tolerance markings

Letter	Tolerance
F	±1%
G	±2%
J	±5%
K	±10%
M	±20%

Example 2.8

A resistor is marked coded with the legend 4R7K. What is its value and tolerance?

Solution

$4.7\,\Omega \pm 10\%$

Example 2.9

A resistor is marked coded with the legend 330RG. What is its value and tolerance?

Solution

$330\,\Omega \pm 2\%$

Example 2.10

A resistor is marked coded with the legend R22M. What is its value and tolerance?

Solution

$0.22\,\Omega \pm 20\%$

Series and parallel combinations of resistors

In order to obtain a particular value of resistance, fixed resistors may be arranged in either series or parallel as shown in Figs 2.11 and 2.12.

The effective resistance of each of the series circuits shown in Fig. 2.11 is simply equal to the sum of the individual resistances. Hence, for Fig. 2.11(a)

$$R = R_1 + R_2$$

while for Fig. 2.11(b)

$$R = R_1 + R_2 + R_3$$

(a)

(b)

Figure 2.11   Resistors in series: (a) two resistors in series (b) three resistors in series

(a)

(b)

Figure 2.12   Resistors in parallel: (a) two resistors in parallel (b) three resistors in parallel

Turning to the parallel resistors shown in Fig. 2.12, the reciprocal of the effective resistance of each circuit is equal to the sum of the reciprocals of the individual resistances. Hence, for Fig. 2.12(a)

$$\frac{1}{R} = \frac{1}{R_1} + \frac{1}{R_2}$$

while for Fig. 2.12(b)

$$\frac{1}{R} = \frac{1}{R_1} + \frac{1}{R_2} + \frac{1}{R_3}$$

In the former case, the formula can be more conveniently re-arranged as follows:

$$R = \frac{R_1 \times R_2}{R_1 + R_2}$$

(You can remember this as the *product* of the two resistance values *divided by* the *sum* of the two resistance values.)

Example 2.11

Resistors of $22\,\Omega$, $47\,\Omega$ and $33\,\Omega$ are connected (a) in series and (b) in parallel. Determine the effective resistance in each case.

Solution

(a) In the series circuit $R = R_1 + R_2 + R_3$, thus

$$R = 22\,\Omega + 47\,\Omega + 33\,\Omega = 102\,\Omega$$

(b) In the parallel circuit:

$$\frac{1}{R} = \frac{1}{R_1} + \frac{1}{R_2} + \frac{1}{R_3}$$

thus

$$\frac{1}{R} = \frac{1}{22\,\Omega} + \frac{1}{47\,\Omega} + \frac{1}{33\,\Omega}$$

or

$$\frac{1}{R} = 0.045 + 0.021 + 0.03$$

thus

$$R = \frac{1}{0.096} = 10.42\,\Omega$$

Example 2.12

Determine the effective resistance of the circuit shown in Fig. 2.13.

Figure 2.13   Circuit for Example 2.12

Solution

The circuit can be progressively simplified as shown in Fig. 2.14. The stages in this simplification are:

(a) R_3 and R_4 are in series and they can be replaced by a single resistance (R_A) of $(12\,\Omega + 27\,\Omega) = 39\,\Omega$.
(b) R_A appears in parallel with R_2. These two resistors can be replaced by a single resistance (R_B) of $(39\,\Omega \times 47\,\Omega)/(39\,\Omega + 47\,\Omega) = 21.3\,\Omega$.
(c) R_B appears in series with R_1. These two resistors can be replaced by a single resistance (R) of $(21.3\,\Omega + 4.7\,\Omega) = 26\,\Omega$.

Example 2.13

A resistance of $50\,\Omega$ 2 W is required. What parallel combination of preferred value resistors will satisfy this requirement? What power rating should each resistor have?

Solution

Two $100\,\Omega$ resistors may be wired in parallel to provide a resistance of $50\,\Omega$ as shown below:

$$R = \frac{R_1 \times R_2}{R_1 + R_2} = \frac{100 \times 100}{100 + 100} = \frac{10\,000}{200} = 50\,\Omega$$

Figure 2.14   Stages in simplifying the circuit of Fig. 2.13

Since the resistors are identical, the applied power will be shared equally between them. Hence each resistor should have a power rating of 1 W.

Resistance and temperature

Figure 2.15 shows how the resistance of a metal conductor (e.g. copper) varies with temperature. Since the resistance of the material increases with temperature, this characteristic is said to exhibit a **positive temperature coefficient (PTC)**. Not all materials have a PTC characteristic. The resistance of a carbon conductor falls with temperature and it is therefore said to exhibit a **negative temperature coefficient (NTC)**.

The resistance of a conductor at a temperature, t, is given by the equation:

$$R_t = R_0(1 + \alpha t + \beta t^2 + \gamma t^3 \ldots)$$

where α, β, γ, etc. are constants and R_0 is the temperature at $0°C$.

The coefficients, β, γ, etc. are quite small and since we are normally only dealing with a relatively restricted temperature range (e.g. $0°C$ to $100°C$) we can usually approximate the characteristic shown in Fig. 2.15 to the straight line law shown in Fig. 2.16. In this case, the equation simplifies to:

$$R_t = R_0(1 + \alpha t)$$

where α is known as the **temperature coefficient** of resistance. Table 2.4 shows some typical values for α (note that α is expressed in $\Omega/\Omega/°C$ or just $/°C$).

Example 2.14

A resistor has a temperature coefficient of $0.001/°C$. If the resistor has a resistance of $1.5\,k\Omega$ at $0°C$, determine its resistance at $80°C$.

Figure 2.15   Variation of resistance with temperature for a metal conductor

Figure 2.16   Straight line approximation of Fig. 2.15

Table 2.4   Temperature coefficient of resistance

Material	Temperature coefficient of resistance, α ($/°C$)
Platinum	+0.0034
Silver	+0.0038
Copper	+0.0043
Iron	+0.0065
Carbon	−0.0005

Solution

Now

$R_t = R_0(1 + \alpha t)$ thus

$R_t = 1.5\,\text{k}\Omega \times (1 + (0.001 \times 80))$

Hence

$R_{80} = 1.5 \times (1 + 0.08) = 1.5 \times 1.08 = 1.62\,\text{k}\Omega$

Example 2.15

A resistor has a temperature coefficient of $0.0005/°C$. If the resistor has a resistance of $680\,\Omega$ at $20°C$, what will its resistance be at $90°C$?

Solution

First we must find the resistance at $0°C$. Rearranging the formula for R_t gives:

$$R_0 = \frac{R_t}{1 + \alpha t} = \frac{680}{1 + (0.0005 \times 20)} = \frac{680}{1 + 0.01}$$

$$= \frac{680}{1.01} = 673.3\,\Omega$$

Now

$R_t = R_0(1 + \alpha t)$ thus

$R_{90} = 680 \times (1 + (0.0005 \times 90))$

Hence

$R_{90} = 680 \times (1 + (0.045)) = 680 \times 1.045$

$\qquad\quad = 650.7\,\Omega$

Example 2.16

A resistor has a resistance of $40\,\Omega$ at $0°C$ and $44\,\Omega$ at $100°C$. Determine the resistor's temperature coefficient.

Solution

First we need to make α the subject of the formula $R_t = R_0(1 + \alpha t)$:

Now

$$1 + \alpha t = \frac{R_t}{R_0} \text{ thus } \alpha t = \frac{R_t}{R_0} - 1$$

Hence

$$\alpha = \frac{1}{t}\left(\frac{R_t}{R_0} - 1\right) = \frac{1}{100}\left(\frac{44}{40} - 1\right)$$

Therefore

$$\alpha = \frac{1}{100}(1.1 - 1) = \frac{1}{100} \times 0.1 = 0.001/°C$$

Thermistors

With conventional resistors we would normally require resistance to remain the same over a wide range of temperatures (i.e. α should be zero). On the other hand, there are applications in which we could use the effect of varying resistance to detect a temperature change. Components that allow us to do this are known as **thermistors**. The resistance of a thermistor changes markedly with temperature and these components are widely used in temperature sensing and temperature compensating applications. Two basic types of thermistor are available, NTC and PTC.

Typical NTC thermistors have resistances which vary from a few hundred (or thousand) ohms at $25°C$ to a few tens (or hundreds) of ohms at $100°C$ (see Fig. 2.17). PTC thermistors, on the

Figure 2.17 Negative temperature coefficient (NTC) thermistor characteristic

other hand, usually have a resistance–temperature characteristic which remains substantially flat (typically at around $100\,\Omega$) over the range $0°C$ to around $75°C$. Above this, and at a critical temperature (usually in the range $80°C$ to $120°C$) their resistance rises very rapidly to values of up to, and beyond, $10\,k\Omega$ (see Fig. 2.18).

A typical application of PTC thermistors is over-current protection. Provided the current passing through the thermistor remains below the threshold current, the effects of self-heating will remain negligible and the resistance of the thermistor will remain low (i.e. approximately the same as the resistance quoted at $25°C$). Under fault conditions, the current exceeds the threshold value by a considerable margin and the thermistor starts to self-heat. The resistance then increases rapidly and, as a consequence, the current falls to the rest value. Typical values of threshold and rest currents are $200\,mA$ and $8\,mA$, respectively, for a

device which exhibits a nominal resistance of $25\,\Omega$ at $25°C$.

Light dependent resistors

Light dependent resistors (LDR) use a semiconductor material (i.e. a material that is neither a conductor nor an insulator) whose electrical characteristics vary according to the amount of incident light. The two semiconductor materials used for the manufacture of LDRs are cadmium sulphide (CdS) and cadmium selenide (CdSe). These materials are most sensitive to light in the visible spectrum, peaking at about $0.6\,\mu m$ for CdS and $0.75\,\mu m$ for CdSe. A typical CdS LDR exhibits a resistance of around $1\,M\Omega$ in complete darkness and less than $1\,k\Omega$ when placed under a bright light source (see Fig. 2.19).

Voltage dependent resistors

The resistance of a voltage dependent resistor (VDR) falls very rapidly when the voltage across it exceeds a nominal value in either direction (see Fig. 2.20). In normal operation, the current flowing in a VDR is negligible, however, when the resistance falls, the current will become appreciable and a significant amount of energy will be absorbed.

VDRs are used as a means of 'clamping' the voltage in a circuit to a pre-determined level. When connected across the supply rails to a circuit (either AC or DC) they are able to offer a measure of protection against supply voltage surges.

Figure 2.18 Positive temperature coefficient (PTC) thermistor characteristic

Figure 2.19 Light dependent resistor (LDR) characteristic

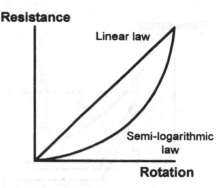

Figure 2.20 Current plotted against voltage for a voltage dependent resistor (VDR) (note that the slope of the graph is inversely proportional to the value of resistance)

Figure 2.22 Characteristics of linear and semi-logarithmic potentiometers

preset resistor are commonly used including open carbon track skeleton presets and fully encapsulated carbon and multi-turn cermet types.

Variable resistors

Variable resistors are available in several including those which use carbon tracks and those which use a wirewound resistance element. In either case, a moving slider makes contact with the resistance element (see Fig. 2.21). Most variable resistors have three (rather than two) terminals and as such are more correctly known as potentiometers.

Carbon potentiometers are available with linear or semi-logarithmic law tracks (see Fig. 2.22) and in rotary or slider formats. Ganged controls, in which several potentiometers are linked together by a common control shaft, are also available.

You will also encounter various forms of preset resistors that are used to make occasional adjustments (e.g. for calibration). Various forms of

Capacitors

A capacitor is a device for storing electric charge. In effect, it is a reservoir into which charge can be deposited and then later extracted. Typical applications include reservoir and smoothing capacitors for use in power supplies, coupling a.c. signals between the stages of amplifiers, and decoupling supply rails (i.e. effectively grounding the supply rails as far as a.c. signals are concerned).

A capacitor need consist of nothing more than two parallel metal plates as shown in Fig. 2.23. If the switch is left open, no charge will appear on the plates and in this condition there will be no electric field in the space between the plates nor any charge stored in the capacitor.

Take a look at the circuit shown in Fig. 2.24(a). With the switch left open, no current will flow and no charge will be present in the capacitor. When the switch is closed (see Fig. 2.24(b)), electrons will be attracted from the positive plate to the positive terminal of the battery. At the same time, a similar number of electrons will move from the negative terminal of the battery to the negative plate. This sudden movement of electrons will manifest itself in a momentary surge of current (conventional current will flow from the positive terminal of the battery towards the positive terminal of the capacitor).

Eventually, enough electrons will have moved to make the e.m.f. between the plates the same as that of the battery. In this state, the capacitor is said to

Figure 2.21 Construction of a carbon track rotary potentiometer

Figure 2.23   Basic parallel plate capacitor

be charged and an electric field will be present in the space between the two plates.

If, at some later time the switch is opened (see Fig. 2.24(c)), the positive plate will be left with a deficiency of electrons while the negative plate will be left with a surplus of electrons. Furthermore, since there is no path for current to flow between the two plates the capacitor will remain charged and a potential difference will be maintained between the plates. In practice, however, the stored charge will slowly decay due to the leakage resistance inside the capacitor.

Capacitance

The unit of capacitance is the farad (F). A capacitor is said to have a capacitance of 1 F if a current of 1 A flows in it when a voltage changing at the rate of 1 V/s is applied to it.

The current flowing in a capacitor will thus be proportional to the product of the capacitance (C) and the rate of change of applied voltage. Hence:

$$i = C \times \text{(rate of change of voltage)}$$

The rate of change of voltage is often represented by the expression dv/dt where dv represents a very small change in voltage and dt represents the corresponding small change in time. Thus:

$$i = C\frac{dv}{dt}$$

Example 2.17

A voltage is changing at a uniform rate from 10 V to 50 V in a period of 0.1 s. If this voltage is applied to a capacitor of 22 μF, determine the current that will flow.

Solution

Now the current flowing will be given by:

$$i = C \times \text{(rate of change of voltage)}$$

thus

$$i = C\left(\frac{\text{change in voltage}}{\text{time}}\right) = 22 \times 10^{-6}\frac{50 - 10}{0.1}$$

thus

$$i = 22 \times 10^{-6} \times \frac{40}{0.1} = 22 \times 400 \times 10^{-6}$$

$$= 8.8 \times 10^{-3} = 8.8\,\text{mA}$$

Charge, capacitance and voltage

The charge or **quantity of electricity** that can be stored in the electric field between the capacitor plates is proportional to the applied voltage and the capacitance of the capacitor. Thus:

Figure 2.24   Effect of applying a voltage to a capacitor: (a) initial state (no charge present); (b) charge rapidly builds up when voltage is applied; (c) charge remains when voltage is removed

$$Q = CV$$

where Q is the charge (in coulombs), C is the capacitance (in farads), and V is the potential difference (in volts).

Example 2.18

A $10\,\mu F$ capacitor is charged to a potential of $250\,V$. Determine the charge stored.

Solution

The charge stored will be given by:

$$Q = CV = 10 \times 10^{-6} \times 250 = 2.5\,mC$$

Energy storage

The energy stored in a capacitor is proportional to the product of the capacitance and the square of the potential difference. Thus:

$$E = 0.5CV^2$$

where E is the energy (in joules), C is the capacitance (in farads), and V is the potential difference (in volts).

Example 2.19

A capacitor of $47\,\mu F$ is required to store an energy of $4\,J$. Determine the potential difference which must be applied.

Solution

The foregoing formula can be re-arranged to make V the subject as follows:

$$V = \left(\frac{E}{0.5C}\right)^{0.5} = \left(\frac{2E}{C}\right)^{0.5} = \left(\frac{2 \times 4}{47 \times 10^{-6}}\right)^{0.5}$$

$$= 130\,V$$

Capacitance and physical characteristics

The capacitance of a capacitor depends upon the physical dimensions of the capacitor (i.e. the size of the plates and the separation between them) and the dielectric material between the plates. The capacitance of a conventional parallel plate capacitor is given by:

$$C = \frac{\varepsilon_0 \varepsilon_r A}{d}$$

where C is the capacitance (in farads), ε_0 is the permittivity of free space, ε_r is the **relative permittivity** of the dielectric medium between the plates), and d is the separation between the plates (in metres).

Example 2.20

A capacitor of $1\,nF$ is required. If a dielectric material of thickness $0.1\,mm$ and relative permittivity 5.4 is available, determine the required plate area.

Solution

Re-arranging the formula $C = \varepsilon_0 \varepsilon_r A/d$ to make A the subject gives:

$$A = \frac{Cd}{\varepsilon_0 \varepsilon_r} = \frac{1 \times 10^{-9} \times 0.1 \times 10^{-3}}{8.854 \times 10^{-12} \times 5.4}$$

$$= \frac{0.1 \times 10^{-12}}{47.8116 \times 10^{-12}}$$

thus

$$A = 0.002\,09\,m^2 \text{ or } 20.9\,cm^2$$

In order to increase the capacitance of a capacitor, many practical components employ multiple plates (see Fig. 2.25). The capacitance is then given by:

$$C = \frac{\varepsilon_0 \varepsilon_r (n - 1) A}{d}$$

where C is the capacitance (in farads), ε_0 is the permittivity of free space, ε_r is the relative permittivity of the dielectric medium between the plates, d is the separation between the plates (in metres) and n is the total number of plates.

Example 2.21

A capacitor consists of six plates each of area $20\,cm^2$ separated by a dielectric of relative

Figure 2.25   Multiple-plate capacitor

permittivity 4.5 and thickness 0.2 mm. Determine the capacitance of the capacitor.

Solution

Using $C = \varepsilon_0 \varepsilon_r (n - 1) A/d$ gives:

$$C = \frac{8.854 \times 10^{-12}(6 - 1)20 \times 10^{-4}}{0.2 \times 10^{-3}}$$

$$= \frac{885.40 \times 10^{-16}}{0.2 \times 10^{-3}}$$

$$C = 4427 \times 10^{-13} = 442.7 \times 10^{-12}\,\text{F}$$

thus $C = 442.7\,\text{pF}$.

Capacitor specifications

The specifications for a capacitor usually include the value of capacitance (expressed in microfarads, nanofarads or picofarads), the voltage rating (i.e. the maximum voltage which can be continuously applied to the capacitor under a given set of conditions), and the accuracy or tolerance (quoted as the maximum permissible percentage deviation from the marked value).

Other practical considerations when selecting capacitors for use in a particular application include temperature coefficient, leakage current, stability and ambient temperature range. Table 2.5 summarizes the properties of five of the most common types of capacitor. Figure 2.26 shows the construction of a typical tubular polystyrene capacitor.

Capacitor markings

The vast majority of capacitors employ written markings which indicate their values, working voltages, and tolerance. The most usual method of marking resin dipped polyester (and other) types of capacitor involves quoting the value (μF, nF or pF), the tolerance (often either 10% or 20%), and the working voltage (often using _ and ~ to indicate d.c. and a.c., respectively). Several manufacturers use two separate lines for their capacitor markings and these have the following meanings:

First line: capacitance (pF or μF) and tolerance (K = 10%, M = 20%)

Second line: rated d.c. voltage and code for the dielectric material

A three-digit code is commonly used to mark monolithic ceramic capacitors. The first two digits correspond to the first two digits of the value while the third digit is a multipler which gives the number of zeros to be added to give the value in picofarads.

Example 2.22

A monolithic ceramic capacitor is marked with the legend '103K'. What is its value?

Table 2.5   Characteristics of common types of capacitor

Parameter	Capacitor type				
	Ceramic	Electrolytic	Metallized film	Mica	Polyester
Capacitance range (F)	2.2 p to 100 n	100 n to 68 m	1 μ to 16 μ	2.2 p to 10 n	10 n to 2.2 μ
Typical tolerance (%)	±10 and ±20	−10 to +50	±20	±1	±20
Typical voltage rating (d.c.)	50 V to 250 V	6.3 V to 400 V	250 V to 600 V	350 V	250 V
Temperature coefficient (ppm/°C)	+100 to −4700	+1000 typical	+100 to 200	+50	+250
Stability	Fair	Poor	Fair	Excellent	Good
Ambient temperature range (°C)	−85 to +85	−40 to +85	−25 to +85	−40 to +85	−40 to +100
Typical applications	Decoupling at high frequency	Smoothing and decoupling at low frequency	Power supplies and power factor correction	Tuned circuits, filters, oscillators	General purpose

Connections to foils made by end caps

Aluminium foil

Plastic film dielectric

Aluminium foil

Figure 2.26   Construction of a typical tubular polystyrene capacitor

1st and 2nd colour bands	
Black	0
Brown	1
Red	2
Orange	3
Yellow	4
Green	5
Blue	6
Violet	7
Grey	8
White	9

3rd coloured band	Multiply by
Black	1
Brown	10
Red	100
Orange	1000
Yellow	10000
Green	100000

4th coloured band (tolerance)	
Red	± 2%
Green	± 5%
White	± 10%
Black	± 20%

5th coloured band (voltage)	
Brown	100V dc
Red	200V dc
Yellow	400V dc

Figure 2.27   Capacitor colour code

Solution

The value (pF) will be given by the first two digits (10) followed by the number of zeros indicated by the third digit (3). The value of the capacitor is thus 10 000 pF or 10 nF. The final letter (K) indicates that the capacitor has a tolerance of 10%.

Some capacitors are marked with coloured stripes to indicate their value and tolerance (see Fig. 2.27).

Example 2.23

A tubular capacitor is marked with the following coloured stripes: brown, green, brown, red, brown. What is its value, tolerance, and working voltage?

Solution

See Fig. 2.28.

Brown, green, brown, red, brown

1st band Brown = 1	2nd band Green = 5	3rd band Brown = x10	4th band Red = ± 2%	5th band Brown = 100V dc

1	5	0	± 2%	100V dc

150pF ± 2% 100V dc

Figure 2.28

Series and parallel combination of capacitors

In order to obtain a particular value of capacitance, fixed capacitors may be arranged in either series or parallel (Figs 2.29 and 2.30).

The reciprocal of the effective capacitance of each of the series circuits shown in Fig. 2.29 is equal to the sum of the reciprocals of the individual capacitances. Hence, for Fig. 2.29(a)

$$\frac{1}{C} = \frac{1}{C_1} + \frac{1}{C_2}$$

while for Fig. 2.29(b)

$$\frac{1}{C} = \frac{1}{C_1} + \frac{1}{C_2} + \frac{1}{C_3}$$

In the former case, the formula can be more conveniently re-arranged as follows:

$$C = \frac{C_1 \times C_2}{C_1 + C_2}$$

(You can remember this as the *product* of the two capacitor values *divided by* the *sum* of the two values.)

For parallel arrangements of capacitors, the effective capacitance of the circuit is simply equal to the sum of the individual capacitances. Hence, for Fig. 2.30(a)

$$C = C_1 + C_2$$

Figure 2.29 Capacitors in series: (a) two capacitors in series; (b) three capacitors in series

while for Fig. 2.30(b)

$$C = C_1 + C_2 + C_3$$

Example 2.24

Determine the effective capacitance of the circuit shown in Fig. 2.31.

Solution

The circuit of Fig. 2.31 can be progressively simplified as shown in Fig. 2.32. The stages in this simplification are:

(a) C_1 and C_2 are in parallel and they can be replaced by a single capacitance (C_A) of $(2\,n + 4\,n) = 6\,n$.

Figure 2.30   Capacitors in parallel: (a) two capacitors in parallel; (b) three capacitors in parallel

Figure 2.31   Circuit for Example 2.24

(b) C_A appears in parallel with C_3. These two capacitors can be replaced by a single capacitance (C_B) of $(6n \times 2n)/(6n + 2n) = 1.5n$.

(c) C_B appears in parallel with C_4. These two capacitors can be replaced by a single capacitance (C) of $(1.5n + 4n) = 5.5n$.

Example 2.25

A capacitance of 50 nF (rated at 100 V) is required. What series combination of preferred value capacitors will satisfy this requirement? What voltage rating should each capacitor have?

Solution

Two 100 μF capacitors wired in series will provide a capacitance of 50 μF, as shown below:

$$C = \frac{C_1 \times C_2}{C_1 + C_2} = \frac{100\,\mu \times 100\,\mu}{100\,\mu + 100\,\mu} = \frac{10\,000}{200} = 50\,\mu F$$

Figure 2.32   Stages in simplifying the circuit of Fig. 2.31

Since the capacitors are of equal value, the applied d.c. potential will be shared equally between them. Thus each capacitor should be rated at 50 V. Note that, in a practical circuit, we could take steps to ensure that the d.c. voltage was shared equally between the two capacitors by wiring equal, high-value (e.g. 100 kΩ) resistors across each capacitor.

Variable capacitors

By moving one set of plates relative to the other, a capacitor can be made variable. The dielectric material used in a variable capacitor can be either air (see Fig. 2.33) or plastic (the latter tend to be more compact). Typical values for variable capacitors tend to range from about 25 pF to 500 pF. These components are commonly used for tuning radio receivers.

Inductors

Inductors provide us with a means of storing electrical energy in the form of a magnetic field. Typical applications include chokes, filters and frequency selective circuits. The electrical characteristics of an inductor are determined by a number of factors including the material of the core (if any), the number of turns, and the physical

Moving plates

Increase
capacitance

Fixed plates

Figure 2.33 Construction of an air-spaced variable capacitor

dimensions of the coil. Figure 2.34 shows the construction of a basic air-cored inductor.

In practice every coil comprises both inductance (L) and resistance (R_S). The circuit of Fig. 2.35 shows these as two discrete components. In reality the inductance and resistance are both distributed throughout the component but it is convenient to treat the inductance and resistance as separate components in the analysis of the circuit.

Take a look at the circuit shown in Fig. 2.36(a). If the switch is left open, no current will flow and

Turns of coiled wire

Connecting wires

Air or high permeability
ferrite core

Figure 2.34 Basic air-cored inductor

Figure 2.35 A practical coil contains inductance **and** resistance

no magnetic flux will be produced by the inductor. If the switch is closed (see Fig. 2.36(b)), current will begin to flow as energy is taken from the supply in order to establish the magnetic field. However, the change in magnetic flux resulting from the appearance of current creates a voltage (an induced e.m.f.) across the coil which opposes the applied e.m.f. from the battery.

The induced e.m.f. results from the changing flux and it effectively prevents an instantaneous rise in current in the circuit. Instead, the current increases slowly to a maximum at a rate which depends upon the ratio of inductance (L) to resistance (R) present in the circuit. After a while, a steady state condition will be reached in which the voltage across the inductor will have decayed to zero and the current will have reached a maximum value (determined by the ratio of V to R, i.e. Ohm's law).

If, after this steady state condition has been achieved, the switch is opened (see Fig. 2.36(c)), the magnetic field will suddenly collapse and the energy will be returned to the circuit in the form of 'back e.m.f.' which will appear across the coil as the field collapses.

Inductance

Inductance is the property of a coil which gives rise to the opposition to a change in the value of

Figure 2.36 Effect of applying a current to an inductor: (a) initial state (no flux present); (b) flux rapidly builds up when current is applied; (c) flux collapses when current is removed

current flowing in it. Any change in the current applied to a coil/inductor will result in an induced voltage appearing across it.

The unit of inductance is the henry (H) and a coil is said to have an inductance of 1 H if a voltage of 1 V is induced across it when a current changing at the rate of 1 A/s is flowing in it.

The voltage induced across the terminals of an inductor will thus be proportional to the product of the inductance (L) and the rate of change of applied current. Hence:

$$e = -L \times \text{(rate of change of current)}$$

(Note that the minus sign indicates the polarity of the voltage, i.e. opposition to the change.)

Note that the rate of change of current is often represented by the expression di/dt where di represents a very small change in current and dt represents the corresponding small change in time.

$$e = -L\frac{di}{dt}$$

Example 2.26

A current increases at a uniform rate from 2 A to 6 A in a period of 250 ms. If this current is applied to an inductor of 600 mH, determine the voltage induced.

Solution

Now the induced voltage will be given by:

$$e = -L \times \text{(rate of change of current)}$$

thus

$$e = -L \times \left(\frac{\text{change in current}}{\text{time}}\right) = \frac{-0.6 \times (6-2)}{0.25}$$

$$= -9.6 \, \text{V}$$

Energy storage

The energy stored in an inductor is proportional to the product of the inductance and the square of the current. Thus:

$$E = 0.5LI^2$$

where E is the energy (in joules), L is the inductance (in henries), and I is the current (in amperes).

Example 2.27

An inductor of 20 mH is required to store an energy of 2.5 J. Determine the current which must be applied.

Solution

The foregoing formula can be re-arranged to make I the subject as follows:

$$I = \left(\frac{E}{0.5L}\right)^{0.5} = \left(\frac{2E}{L}\right)^{0.5} = \left(\frac{2 \times 2.5}{20 \times 10^{-3}}\right)^{0.5} = \sqrt{250}$$

$$= 15.811 \, \text{A}$$

Inductance and physical characteristics

The inductance of an inductor depends upon the physical dimensions of the inductor (e.g. the length and diameter of the winding), the number of turns, and the permeability of the material of the core. The inductance of an inductor is given by:

$$L = \frac{\mu_0 \mu_r n^2 A}{l}$$

where L is the inductance (in henries), μ_0 is the permeability of free space, μ_r is the relative permeability of the magnetic core, l is the length of the core (in metres), and A is the cross-sectional area of the core (in square metres).

Example 2.28

An inductor of 100 mH is required. If a closed magnetic core of length 20 cm, cross-sectional area 15 cm^2 and relative permeability 500 is available, determine the number of turns required.

Solution

Re-arranging the formula $L = \mu_0 \mu_r n^2 A/l$ to make n the subject gives:

$$n = \frac{L \times l}{\mu_0 \mu_r n^2 A} = \frac{0.1 \times 0.02}{12.57 \times 10^{-7} \times 20 \times 10^{-4}}$$

$$= \frac{0.002}{251.4 \times 10^{-11}}$$

thus

$$n = \frac{200}{251.4} \times 10^6 = 0.892 \times 10^3 = 892 \, \text{turns}$$

Inductor specifications

Inductor specifications normally include the value of inductance (expressed in henries, millihenries or microhenries), the current rating (i.e. the maximum current which can be continuously applied to the inductor under a given set of conditions), and the accuracy or tolerance (quoted as the maximum permissible percentage deviation from the marked value). Other considerations may include the temperature coefficient of the inductance (usually expressed in parts per million, ppm, per unit temperature change), the stability of the inductor, the d.c. resistance of the coil windings (ideally zero), the Q-factor (quality factor) of the coil, and the recommended working frequency range. Table 2.6 summarizes the properties of four common types of inductor. Figure 2.37 shows the construction of a typical ferrite-cored inductor.

Inductor markings

As with capacitors, the vast majority of inductors use written markings to indicate values, working current, and tolerance. Some small inductors are marked with coloured stripes to indicate their value and tolerance (the standard colour values are used and inductance is normally expressed in microhenries).

Series and parallel combinations of inductors

In order to obtain a particular value of inductance, fixed inductors may be arranged in either series or

Figure 2.37   Construction of a typical ferrite cored inductor

parallel as shown in Figs 2.38 and 2.39. The effective inductance of each of the series circuits shown in Fig. 2.38 is simply equal to the sum of the individual inductances. Hence, for Fig. 2.38(a)

$$L = L_1 + L_2$$

while for Fig. 2.38(b)

$$L = L_1 + L_2 + L_3$$

Figure 2.38   Inductors in series: (a) two inductors in series (b) three inductors in series

Table 2.6   Characteristics of common types of inductor

Parameter	Inductor type			
	Air cored	Ferrite cored	Ferrite pot cored	Iron cored
Core material	Air	Ferrite	Ferrite	Iron
Inductance range (H)	50 n to 100 μ	10 μ to 1 m	1 m to 100 m	20 m to 20 H
Typical d.c. resistance (Ω)	0.05 to 10	1 to 100	2 to 100	10 to 200
Typical tolerance (%)	±10	±10	±10	±10
Typical Q-factor	60	80	40	20
Typical frequency range (Hz)	1 M to 500 M	100 k to 100 M	1 k to 10 M	50 Hz to 10 kHz
Typical applications	Tuned circuits	Filters and HF transformers	LF and MF filters and transformers	Smoothing chokes and LF filters

(a)

(b)

Figure 2.39 Inductors in parallel: (a) two inductors in parallel (b) three inductors in parallel

Turning to the parallel inductors shown in Fig. 2.39, the reciprocal of the effective inductance of each circuit is equal to the sum of the reciprocals of the individual inductances. Hence, for Fig. 2.39(a)

$$\frac{1}{L} = \frac{1}{L_1} + \frac{1}{L_2}$$

while for Fig. 2.39(b)

$$\frac{1}{L} = \frac{1}{L_1} + \frac{1}{L_2} + \frac{1}{L_3}$$

In the former case, the formula can be more conveniently re-arranged as follows:

$$L = \frac{L_1 \times L_2}{L_1 + L_2}$$

(You can remember this as the *product* of the two inductance values *divided by* the *sum* of the two values.)

Example 2.29

An inductance of 5 mH (rated at 2 A) is required. What parallel combination of preferred value inductors will satisfy this requirement?

Solution

Two 10 mH inductors may be wired in parallel to provide an inductance of 5 mH as shown below:

$$L = \frac{L_1 \times L_2}{L_1 + L_2} = \frac{10\,\text{m} \times 10\,\text{m}}{10\,\text{m} + 10\,\text{m}} = \frac{100\,\text{m}}{20\,\text{m}} = 5\,\text{mH}$$

Since the inductors are identical, the applied current will be shared equally between them. Hence each inductor should have a current rating of 1 A.

Example 2.30

Determine the effective inductance of the circuit shown in Fig. 2.40.

Solution

The circuit can be progressively simplified as shown in Fig. 2.41. The stages in this simplification are:

(a) L_1 and L_2 are in series and they can be replaced by a single inductance (L_A) of (60 m + 60 m) = 120 mH.
(b) L_A appears in parallel with L_3. These two inductors can be replaced by a single inductance (L_B) of (120 m × 120 m)/(120 m + 120 m) = 60 mH.
(c) L_B appears in series with L_4. These two inductors can be replaced by a single inductance (L) of (60 m + 50 m) = 110 mH.

Variable inductors

A ferrite cored inductor can be made variable by moving its core in or out of the former onto which the coil is wound. Many small inductors have threaded ferrite cores to make this possible. Such

Figure 2.40 Circuit for Example 2.29

Figure 2.41 Stages in simplifying the circuit of Fig. 2.40

inductors are often used in radio and high-frequency applications where precise tuning is required.

Formulae introduced in this chapter

Component tolerance:
(page 18)

$$\text{Tolerance} = \frac{\text{error}}{\text{marked value}} \times 100\%$$

Resistors in series:
(page 25)

$$R = R_1 + R_2 + R_3$$

Resistors in parallel:
(page 25)

$$\frac{1}{R} = \frac{1}{R_1} + \frac{1}{R_2} + \frac{1}{R_3}$$

Two resistors in parallel:
(page 25)

$$R = \frac{R_1 \times R_2}{R_1 + R_2}$$

Resistance and temperature:
(page 26)

$$R_t = R_0(1 + \alpha t)$$

Current flowing in a capacitor:
(page 30)

$$i = C\frac{\mathrm{d}v}{\mathrm{d}t}$$

Charge stored in a capacitor:
(page 32)

$$Q = CV$$

Energy stored in a capacitor:
(page 32)

$$E = \frac{1}{2}CV^2$$

Capacitance of a capacitor:
(page 32)

$$C = \frac{\varepsilon_0 \varepsilon_r A}{d}$$

Capacitors in series:
(page 35)

$$\frac{1}{C} = \frac{1}{C_1} + \frac{1}{C_2} + \frac{1}{C_3}$$

Two capacitors in series:
(page 35)

$$C = \frac{C_1 \times C_2}{C_1 + C_2}$$

Capacitors in parallel:
(page 36)

$$C = C_1 + C_2 + C_3$$

Induced e.m.f. in an inductor:
(page 39)

$$e = -L\frac{\mathrm{d}i}{\mathrm{d}t}$$

Energy stored in an inductor:
(page 39)

$$E = \frac{1}{2}LI^2$$

Inductance of an inductor:
(page 39)

$$L = \frac{\mu_0 \mu_r n^2 A}{l}$$

Inductors in series:
(page 40)

$$L = L_1 + L_2 + L_3$$

Inductors in parallel:
(page 41)

$$\frac{1}{L} = \frac{1}{L_1} + \frac{1}{L_2} + \frac{1}{L_3}$$

Two inductors in parallel:
(page 41)

$$L = \frac{L_1 \times L_2}{L_1 + L_2}$$

Circuit symbols introduced in this chapter

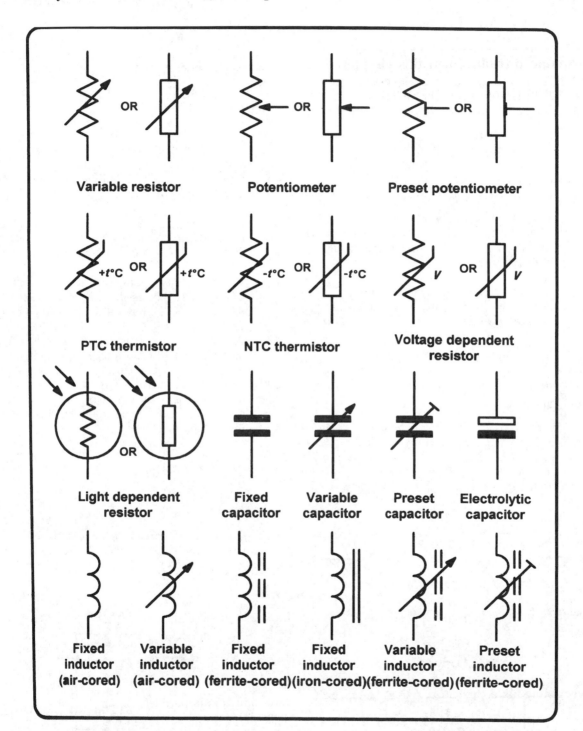

Figure 2.42

Problems

2.1   A power supply rated at 15 V, 0.25 A is to be tested at full rated output. What value of load resistance is required and what power rating should it have? What type of resistor is most suitable for this application and why?

2.2   Determine the value and tolerance of resistors marked with the following coloured bands:

(a) red, violet, yellow, gold;
(b) brown, black, black, silver;
(c) blue, grey, green, gold;
(d) orange, white, silver, gold;
(e) red, red, black, brown, red.

2.3   A batch of resistors are all marked yellow, violet, black, gold. If a resistor is selected from this batch within what range would you expect its value to be?

2.4   Resistors of 27 Ω, 33 Ω, 56 Ω and 68 Ω are available. How can two or more of these resistors be arranged to realize the following resistance values:

(a) 60 Ω
(b) 14.9 Ω
(c) 124 Ω
(d) 11.7 Ω
(e) 128 Ω.

2.5   Three 100 Ω resistors are connected as shown in Fig. 2.43. Determine the effective resistance of the circuit.

2.6   Determine the effective resistance of the circuit shown in Fig. 2.44.

Figure 2.43

Figure 2.44

Figure 2.45

2.7   Determine the resistance of the network shown in Fig. 2.45 looking into terminals A and B with (a) terminals C and D open-circuit and (b) terminals C and D short-circuit.

2.8   A resistor has a temperature coefficient of 0.0008/°C. If the resistor has a resistance of 390 Ω at 0°C, determine its resistance at 55°C.

2.9   A resistor has a temperature coefficient of 0.004/°C. If the resistor has a resistance of 82 kΩ at 20°C, what will its resistance be at 75°C?

2.10 A resistor has a resistance of 218 Ω at 0°C and 225 Ω at 100°C. Determine the resistor's temperature coefficient.

2.11 Capacitors of 1 μF, 3.3 μF, 4.7 μF and 10 μF are available. How can two or more of these capacitors be arranged to realize the following capacitance values:

(a) 8 μF
(b) 11 μF
(c) 19 μF
(d) 0.91 μF
(e) 1.94 μF.

2.12 Three 180 pF capacitors are connected (a) in series and (b) in parallel. Determine the effective capacitance in each case.

Figure 2.46

2.13   Determine the effective capacitance of the circuit shown in Fig. 2.46.

2.14   A capacitor of 330 μF is charged to a potential of 63 V. Determine the quantity of charge stored.

2.15   A parallel plate capacitor has plates of area 0.02 m². Determine the capacitance of the capacitor if the plates are separated by a dielectric of thickness 0.5 mm and relative permittivity 5.6.

2.16   A capacitor is required to store 0.5 J of energy when charged from a 120 V d.c. supply. Determine the value of capacitor required.

2.17   The current in a 2.5 H inductor increases uniformly from zero to 50 mA in 400 ms. Determine the induced e.m.f.

2.18   An inductor comprises 200 turns of wire wound on a closed magnetic core of length 24 cm, cross-sectional area 10 cm² and relative permeability 650. Determine the inductance of the inductor.

2.19   A current of 4 A flows in a 60 mH inductor. Determine the energy stored.

2.20   Inductors of 10 mH, 22 mH, 60 mH and 100 mH are available. How can two or more of these inductors be arranged to realize the following inductance values:

(a)   6.2 mH
(b)   6.9 mH
(c)   32 mH
(d)   70 mH
(e)   170 mH.

(Answers to these problems appear on page 260.)

3

D.C. circuits

In many cases, Ohm's law alone is insufficient to determine the magnitude of the voltages and currents present in a circuit. This chapter introduces several techniques that simplify the task of solving complex circuits. It also introduces the concept of exponential growth and decay of voltage and current in circuits containing capacitance and resistance and inductance and resistance. It concludes by showing how humble C–R and L–R circuits can be used for shaping the waveforms found in electronic circuits. We start by introducing two of the most useful laws of electronics.

Kirchhoff's laws

Kirchhoff's laws relate to the algebraic sum of currents at a junction (or node) or voltages in a network (or mesh). The term 'algebraic' simply indicates that the polarity of each current or voltage drop must be taken into account by giving it an appropriate sign, either positive (+) or negative (−).

Kirchhoff's current law states that the algebraic sum of the currents present at a junction (node) in a circuit is zero (see Fig. 3.1).

Kirchhoff's voltage law states that the algebraic sum of the potential drops in a closed network (or 'mesh') is zero (see Fig. 3.2).

Example 3.1

Determine the currents and voltages in the circuit of Fig. 3.3.

Solution

In order to solve the circuit shown in Fig. 3.3, it is first necessary to identify the currents and voltages as shown in Figs 3.4 and 3.5.

By applying Kirchhoff's current law at node A in Fig. 3.4:

$$I_1 + I_2 = I_3 \tag{1}$$

By applying Kirchhoff's voltage law in loop A of Fig. 3.5:

$$12 = V_1 + V_3 \tag{2}$$

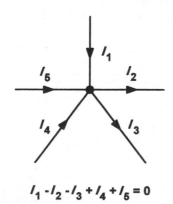

$$I_1 - I_2 - I_3 + I_4 + I_5 = 0$$

Convention:
Currents flowing towards the junction are positive (+)
Currents flowing away from the junction are negative (-)

Figure 3.1 Kirchhoff's current law

$$E_1 - V_1 - V_2 - E_2 = 0$$

Convention:
Move clockwise around the circuit starting with the positive terminal of the largest EMF.
Voltages acting in the same sense are positive (+)
Voltages acting in the opposite sense are negative (-)

Figure 3.2 Kirchhoff's voltage law

Figure 3.3

Figure 3.4 Currents in Example 3.1

Figure 3.5 Voltages in Example 3.1

By applying Kirchhoff's voltage law in loop B of Fig. 3.5:

$$9 = V_2 + V_3 \qquad (3)$$

By applying Ohm's law:

$$V_1 = I_1 R_1 = I_1 \times 68 \qquad (4)$$

$$V_2 = I_2 R_2 = I_2 \times 33 \qquad (5)$$

and

$$V_3 = I_3 R_3 = I_3 \times 22 \qquad (6)$$

From (1) and (6):

$$V_3 = (I_1 + I_2) \times 22 \qquad (7)$$

Combining (4), (7) and (2):

$$12 = I_1 \times 68 + (I_1 + I_2) \times 22$$

or

$$12 = 68I_1 + 22I_1 + 22I_2$$

thus

$$12 = 90I_1 + 22I_2 \qquad (8)$$

Combining (4), (7) and (3):

$$9 = I_2 \times 33 + (I_1 + I_2) \times 22 \qquad (9)$$

or

$$9 = 33I_2 + 22I_1 + 22I_2$$

thus

$$9 = 22I_1 + 55I_2 \qquad (10)$$

Multiplying (8) by 5 gives:

$$60 = 450I_1 + 110I_2 \qquad (11)$$

Multiplying (10) by 2 gives:

$$18 = 44I_1 + 110I_2 \qquad (12)$$

Subtracting (12) from (11):

$$60 - 18 = 450I_1 - 44I_1$$

$$42 = 406I_1$$

thus

$$I_1 = 42/406 = 0.103 \, \text{A}$$

From (8):

$$12 = 90 \times 0.103 + 22I_2$$

thus

$$12 = 9.27 + 22I_2$$

or

$$2.73 = 22I_2$$

thus

$$I_2 = 2.73/22 = 0.124 \, \text{A}$$

Figure 3.6   Potential divider circuit

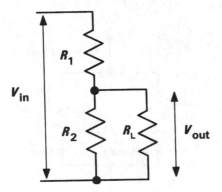

Figure 3.7   Loaded potential divider

Figure 3.8

From (1):

$$I_3 = I_1 + I_2$$

thus

$$I_3 = 0.103 + 0.124 = 0.227 \, \text{A}$$

From (4):

$$V_1 = 0.103 \times 68 = 7 \, \text{V}$$

From (5):

$$V_2 = 0.124 \times 33 = 4 \, \text{V}$$

From (6):

$$V_3 = 0.227 \times 22 = 5 \, \text{V}$$

Check from (2):

$$12 = V_1 + V_3 = 7 + 5 = 12$$

Check from (3):

$$9 = V_2 + V_3 = 4 + 5 = 9$$

The potential divider

The potential divider circuit (see Fig. 3.6) is commonly used to reduce voltage levels in a circuit. The output voltage produced by the circuit is given by:

$$V_{\text{out}} = V_{\text{in}} \times \frac{R_2}{R_1 + R_2}$$

It is, however, important to note that the output voltage (V_{out}) will fall when current is drawn from the arrangement. Figure 3.7 shows the effect of **loading** the potential divider circuit.

In the loaded potential divider (Fig. 3.7) the output voltage is given by:

$$V_{\text{out}} = V_{\text{in}} \times \frac{R_{\text{p}}}{R_1 + R_{\text{p}}}$$

where

$$R_{\text{p}} = \frac{(R_2 \times R_{\text{L}})}{(R_2 + R_{\text{L}})}$$

Example 3.2

The potential divider shown in Fig. 3.8 is used as a simple voltage calibrator. Determine the output voltage produced by the circuit:

(a) when the output terminals are left open-circuit (i.e. no load is connected); and
(b) when the output is loaded by a resistance of 10 kΩ.

Solution

(a) In the first case we can simply apply the formula:

$$V_{\text{out}} = V_{\text{in}} \times \frac{R_2}{R_1 + R_2}$$

where $V_{\text{in}} = 5\,\text{V}$, $R_1 = 4\,\text{k}\Omega$ and $R_2 = 1\,\text{k}\Omega$. Hence

$$V_{\text{out}} = 5 \times \frac{1}{4 + 1} = 5 \times \frac{1}{5} = 1\,\text{V}$$

(b) In the second case we need to take into account the effect of the $10\,\text{k}\Omega$ resistor connected to the output terminals of the potential divider.

First we need to find the equivalent resistance of the parallel combination of R_2 and R_L:

$$R_p = \frac{(R_2 \times R_L)}{(R_2 + R_L)} = \frac{(1\,\text{k}\Omega \times 10\,\text{k}\Omega)}{(1\,\text{k}\Omega + 10\,\text{k}\Omega)}$$

$$= 10/11 = 0.909\,\text{k}\Omega$$

Then we can determine the output voltage from:

$$V_{\text{out}} = V_{\text{in}} \times \frac{R_p}{R_1 + R_p} = 5 \times \frac{0.909}{4 + 0.909}$$

$$= 5 \times \frac{0.909}{4.909}$$

Hence

$$V_{\text{out}} = 5 \times 0.185 = 0.925\,\text{V}$$

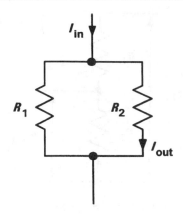

Figure 3.9 Current divider circuit

Figure 3.10

The current divider

The current divider circuit (see Fig. 3.9) is used to divert current from one branch of a circuit to another. The output current produced by the circuit is given by:

$$I_{\text{out}} = I_{\text{in}} \times \frac{R_1}{R_1 + R_2}$$

It is, however, important to note that the output current (I_{out}) will fall when the load connected to the output terminals has any appreciable resistance.

Example 3.3

A moving coil meter requires a current of 1 mA to provide full-scale deflection. If the meter coil has a resistance of $100\,\Omega$ and is to be used as a milliammeter reading 5 mA full-scale, determine the value of parallel shunt resistor required.

Solution

This problem may sound a little complicated so it is worth taking a look at the equivalent circuit of the meter (Fig. 3.10) and comparing it with the current divider shown in Fig. 3.9.

We can apply the current divider formula, replacing I_{out} with I_{m} (the meter full-scale deflection current) and R_2 with R_{m} (the meter resistance). R_1 is the required value of shunt resistor, R_{s}. Hence:

$$I_{\text{m}} = I_{\text{in}} \times \frac{R_{\text{s}}}{R_{\text{s}} + R_{\text{m}}}$$

Re-arranging the formula gives:

$$I_{\text{m}} \times (R_{\text{s}} + R_{\text{m}}) = I_{\text{in}} \times R_{\text{s}}$$

thus

$$I_{\text{m}} R_{\text{s}} + I_{\text{m}} R_{\text{m}} = I_{\text{in}} R_{\text{s}}$$

or

$$I_{\text{in}} R_{\text{s}} - I_{\text{m}} R_{\text{s}} = I_{\text{m}} R_{\text{m}}$$

Figure 3.11   Basic Wheatstone bridge arrangement

hence

$$R_s(I_{in} - I_m) = I_m R_m$$

and

$$R_s = \frac{I_m R_m}{(I_{in} - I_m)}$$

Now $I_m = 1\,\text{mA}$, $R_m = 100\,\Omega$, $I_{in} = 5\,\text{mA}$, thus

$$R_s = \frac{1\,\text{mA} \times 100\,\Omega}{(5\,\text{mA} - 1\,\text{mA})} = \frac{1}{4} \times 100\,\Omega = 25\,\Omega$$

The Wheatstone bridge

The Wheatstone bridge forms the basis of a number of electronic circuits including several that are used in instrumentation and measurement. The basic form of Wheatstone bridge is shown in Fig. 3.11. The voltage developed between A and B will be zero when the voltage between A and Y is the same as that between B and Y. In effect, $R1$ and $R2$ constitute a potential divider as do $R3$ and $R4$. The bridge will be **balanced** (and $V_{AB} = 0$) when the ratio of $R1{:}R2$ is the same as the ratio $R3{:}R4$. Hence, at balance:

$$R1/R2 = R3/R4$$

A practical form of Wheatstone bridge that can be used for measuring unknown resistances is shown in Fig. 3.12. $R1$ and $R2$ constitute the two **ratio arms** while one arm (that occupied by $R3$ in Fig. 3.11) is replaced by a calibrated variable resistor. The unknown resistor, R_x, is connected in the fourth arm.

At balance:

$$R1/R2 = RV/R_x \quad \text{thus} \quad R_x = R2/R1 \times RV$$

Figure 3.12   Practical Wheatstone bridge

Example 3.4

A Wheatstone bridge is based on the circuit shown in Fig. 3.12. If $R1$ and $R2$ can each be switched so that they have values of either $100\,\Omega$ or $1\,\text{k}\Omega$ and RV is variable between $10\,\Omega$ and $10\,\text{k}\Omega$, determine the range of resistance values that can be measured.

Solution

The maximum value of resistance that can be measured will correspond to the largest ratio of $R2{:}R1$ (i.e. when $R2$ is $1\,\text{k}\Omega$ and $R1$ is $100\,\Omega$) and the highest value of RV (i.e. $10\,\text{k}\Omega$). In this case:

$$R_x = R2/R1 \times RV = 1\,\text{k}\Omega/100\,\Omega \times 10\,\text{k}\Omega$$
$$= 10 \times 10\,\text{k}\Omega = 100\,\text{k}\Omega$$

The minimum value of resistance that can be measured will correspond to the smallest ratio of $R2{:}R1$ (i.e. when $R2$ is $100\,\Omega$ and $R1$ is $1\,\text{k}\Omega$) and the smallest value of RV (i.e. $10\,\Omega$). In this case:

$$R_x = R2/R1 \times RV = 100\,\Omega/1\,\text{k}\Omega \times 10\,\Omega$$
$$= 0.1 \times 10\,\Omega = 1\,\Omega$$

Hence the range of values that can be measured extends from $1\,\Omega$ to $100\,\text{k}\Omega$.

Thévenin's theorem

Thévenin's theorem allows us to replace a complicated network of resistances and voltage sources with a simple equivalent circuit comprising a single **voltage source** connected in series with a single resistance (see Fig. 3.13).

The single voltage source in the Thévenin equivalent circuit, V_{OC}, is simply the voltage that

Figure 3.13 Thévenin equivalent circuit

Figure 3.14

Figure 3.15 Determining the equivalent resistance in Fig. 3.14

appears between the terminals when nothing is connected to it. In other words, it is the *open-circuit* voltage that would appear between X and Y.

The single resistance that appears in the Thévenin equivalent circuit, R, is the resistance that would be seen *looking into* the network between X and Y when all of the voltage sources (assumed perfect) are replaced by *short-circuit* connections. Note that if the voltage sources are not perfect (i.e. if they have some internal resistance) the equivalent circuit must be constructed on the basis that each voltage source is replaced by its own internal resistance.

Once we have values for V_{OC} and R, we can determine how the network will behave when it is connected to a load (i.e. when a resistor is connected across the terminals X and Y).

Example 3.5

Figure 3.14 shows a Wheatstone bridge. Determine the current that will flow in a $100\,\Omega$ load connected between terminals A and B.

Solution

First we need to find the Thévenin equivalent of the circuit. To find V_{OC} we can treat the bridge arrangement as two potential dividers. The voltage across $R2$ and $R4$ will be given by:

For $R2$

$$V = 10 \times \frac{R2}{R1 + R2} = 10 \times \frac{600}{500 + 600}$$
$$= 10 \times 0.5454 = 5.454\,\text{V}$$

Hence the voltage at A, relative to Y, will be 5.454 V.

For $R4$

$$V = 10 \times \frac{R4}{R3 + R4} = 10 \times \frac{400}{500 + 400}$$
$$= 10 \times 0.4444 = 4.444\,\text{V}$$

Hence the voltage at B, relative to Y, will be 4.444 V.

The voltage V_{AB} will be the difference between V_{AY} and V_{BY}, hence the open-circuit output voltage, V_{AB}, will be given by:

$$V_{AB} = V_{AY} - V_{BY} = 5.454 - 4.444 = 1.01\,\text{V}$$

Next we need to find the Thévenin equivalent resistance. To do this, we can redraw the circuit, replacing the battery with a short-circuit, as shown in Fig. 3.15.

The equivalent resistance is given by:

$$R = \frac{R1 \times R2}{R1 + R2} + \frac{R3 \times R4}{R3 + R4}$$
$$= \frac{500 \times 600}{500 + 600} + \frac{500 \times 400}{500 + 400}$$

thus

$$R = \frac{300\,000}{1100} + \frac{200\,000}{900} = 272.7 + 222.2$$
$$= 494.9\,\Omega$$

Figure 3.16   Thévenin equivalent of the circuit in Fig. 3.14

Figure 3.17   Determining the current when the Thévenin equivalent circuit is loaded

Figure 3.18   Norton equivalent circuit

Figure 3.19

The Thévenin equivalent circuit is shown in Fig. 3.16. To determine the current in a $100\,\Omega$ load connected between A and B, we can make use of the Thévenin equivalent circuit by simply adding a $100\,\Omega$ resistor to the circuit and applying Ohm's law, as shown in Fig. 3.17.

The current flowing in Fig. 3.17 will be given by:

$$I = \frac{V}{R+100} + \frac{10}{494.9+100} = \frac{10}{594.9} = 0.0168\,\text{A}$$
$$= 16.8\,\text{mA}$$

Norton's theorem

Norton's theorem provides an alternative method of reducing a complex network to a simple equivalent circuit. Unlike Thévenin's theorem, Norton's theorem makes use of a **current source** rather than a voltage source. The Norton equivalent circuit allows us to replace a complicated network of resistances and voltage sources with a simple equivalent circuit comprising a single constant current source connected in parallel with a single resistance (see Fig. 3.18).

The constant current source in the Norton equivalent circuit, I_{SC}, is simply the *short-circuit* current that would flow if X and Y were to be linked directly together.

The resistance that appears in the Norton equivalent circuit, R, is the resistance that would be seen *looking into* the network between X and Y when all of the voltage sources are replaced by *short-circuit* connections. Once again, it is worth noting that, if the voltage sources have any appreciable internal resistance, the equivalent circuit must be constructed on the basis that each voltage source is replaced by its own internal resistance.

As with the Thévenin equivalent, we can determine how a network will behave by obtaining values for I_{SC} and R.

Example 3.6

Three parallel connected temperature sensors having the following characteristics are connected in parallel as shown in Fig. 3.19.

Sensor	A	B	C
Output voltage (open circuit)	20 mV	30 mV	10 mV
Internal resistance	5 kΩ	3 kΩ	2 kΩ

Figure 3.20 Determining the equivalent resistance in Fig. 3.19

Figure 3.21 Norton equivalent of the circuit in Fig. 3.19

Figure 3.22 Determining the voltage when the Norton equivalent circuit is loaded

Determine the voltage produced when the arrangement is connected to a moving-coil meter having a resistance of $1\,k\Omega$.

Solution

First we need to find the Norton equivalent of the circuit. To find I_{SC} we can determine the short-circuit current from each sensor and add them together.

For sensor A

$$I = \frac{V}{R} + \frac{20\,mV}{5\,k\Omega} = 4\,\mu A$$

For sensor B

$$I = \frac{V}{R} + \frac{30\,mV}{3\,k\Omega} = 10\,\mu A$$

For sensor C

$$I = \frac{V}{R} + \frac{10\,mV}{2\,k\Omega} = 5\,\mu A$$

The total current, I_{SC}, will be given by:

$$I_{SC} = 4\,\mu A + 10\,\mu A + 5\,\mu A = 19\,\mu A$$

Next we need to find the Norton equivalent resistance. To do this, we can redraw the circuit showing each sensor replaced by its internal resistance, as shown in Fig. 3.20.

The equivalent resistance is given by:

$$\frac{1}{R} = \frac{1}{R1} + \frac{1}{R2} + \frac{1}{R3} = \frac{1}{5\,k\Omega} + \frac{1}{3\,k\Omega} + \frac{1}{2\,k\Omega}$$

$$= 0.0002 + 0.000\,33 + 0.0005$$

thus

$$\frac{1}{R} = 0.001\,03 \text{ or } R = 968\,\Omega$$

The Norton equivalent circuit is shown in Fig. 3.21. To determine the voltage in a $1\,k\Omega$ moving-coil meter connected between A and B, we can make use of the Norton equivalent circuit by simply adding a $1\,k\Omega$ resistor to the circuit and applying Ohm's law, as shown in Fig. 3.22.

The voltage appearing across the moving-coil meter in Fig. 3.22 will be given by:

$$V = I_{SC} \times \frac{R \times R_M}{R + R_M} = 19\,\mu A \times \frac{1000 \times 968}{1000 + 968}$$

$$= 19\,\mu A \times 492\,\Omega$$

or $V = 9.35\,mV$.

C–R circuits

Networks of capacitors and resistors (known as C–R circuits) form the basis of many timing and pulse shaping circuits and are thus often found in practical electronic circuits.

Charging

A simple C–R circuit is shown in Fig. 3.23. When the network is connected to a constant voltage source (V_s), as shown in Fig. 3.24, the voltage (v_c) across the (initially uncharged) capacitor voltage will rise exponentially as shown in Fig. 3.25. At the same time, the current in the circuit (i) will fall, as shown in Fig. 3.26.

The rate of growth of voltage with time (and decay of current with time) will be dependent upon the

Figure 3.23   Simple *C–R* circuit

Figure 3.24   *C–R* circuit (*C* charging through *R*)

Figure 3.25   Exponential growth of capacitor voltage (v_C) in Fig. 3.24

Figure 3.26   Exponential decay of current (*i*) in Fig. 3.24

product of capacitance and resistance. This value is known as the **time constant** of the circuit. Hence:

time constant, $t = C \times R$

where C is the value of capacitance (F), R is the resistance (Ω), and t is the time constant (s).

The voltage developed across the charging capacitor (v_c) varies with time (t) according to the relationship:

$$v_c = V_s(1 - e^{-t/CR})$$

where v_c is the capacitor voltage, V_s is the d.c. supply voltage, t is the time, and CR is the time constant of the circuit (equal to the product of capacitance, C, and resistance, R).

The capacitor voltage will rise to approximately 63% of the supply voltage in a time interval equal to the time constant. At the end of the next interval of time equal to the time constant (i.e. after an elapsed time equal to $2CR$) the voltage will have risen by 63% of the remainder, and so on.

In theory, the capacitor will **never** become fully charged. However, after a period of time equal to $5CR$, the capacitor voltage will to all intents and purposes be equal to the supply voltage. At this point the capacitor voltage will have risen to 99.3% of its final value and we can consider it to be fully charged.

During charging, the current in the capacitor (i) varies with time (t) according to the relationship:

$$i = V_s e^{-t/CR}$$

where v_c is the capacitor voltage, V_s is the supply voltage, t is the time, C is the capacitance, and R is the resistance.

The current will fall to approximately 37% of the initial current in a time equal to the time constant. At the end of the next interval of time equal to the time constant (i.e. after a total time of $2CR$ has elapsed) the current will have fallen by a further 37% of the remainder, and so on.

Example 3.7

An initially uncharged capacitor of $1\,\mu F$ is charged from a 9 V d.c. supply via a $3.3\,M\Omega$ resistor. Determine the capacitor voltage 1 s after connecting the supply.

Solution

The formula for exponential growth of voltage in the capacitor is:

$$v_c = V_s(1 - e^{-t/CR})$$

Figure 3.27   *C–R* circuit (*C* discharges through *R*)

where $V_s = 9\,V$, $t = 1\,s$ and $CR = 1\,\mu F \times 3.3\,M\Omega = 3.3\,s$. Thus

$$v_c = 9(1 - e^{-1/3.3})$$

or

$$v_c = 9(1 - 0.738)$$

hence

$$v_c = 9 \times 0.262 = 2.358\,V$$

Discharge

Having considered the situation when a capacitor is being charged, let's consider what happens when an already charged capacitor is discharged. When the fully charged capacitor from Fig. 3.24 is connected as shown in Fig. 3.27, the capacitor will discharge through the resistor, and the capacitor voltage (v_c) will fall exponentially with time, as shown in Fig. 3.28. The current in the circuit (i) will also fall, as shown in Fig. 3.29. The rate of discharge (i.e. the rate of decay of voltage with time) will once again be governed by the time constant of the circuit ($C \times R$).

The voltage developed across the discharging capacitor (v_c) varies with time (t) according to the relationship:

$$v_c = V_s e^{-t/CR}$$

The capacitor voltage will fall to approximately 37% of the initial voltage in a time equal to the time constant. At the end of the next interval of time equal to the time constant (i.e. after an elapsed time equal to $2CR$) the voltage will have fallen by 37% of the remainder, and so on.

In theory, the capacitor will **never** become fully discharged. After a period of time equal to $5CR$, however, the capacitor voltage will to all intents and purposes be zero. At this point the capacitor voltage will have fallen below 1% of its initial value. At this point we can consider it to be fully discharged.

Figure 3.28   Exponential decay of capacitor voltage (v_c) in Fig. 3.27

Figure 3.29   Exponential decay of current (i) in Fig. 3.27

As with charging, the current in the capacitor (i) varies with time (t) according to the relationship:

$$i = V_s\,e^{-t/CR}$$

where v_c is the capacitor voltage, V_s is the supply voltage, t is the time, C is the capacitance, and R is the resistance.

The current will fall to approximately 37% of the initial voltage in a time equal to the time constant. At the end of the next interval of time equal to the time constant (i.e. after a total time of $2CR$ has elapsed) the voltage will have fallen by a further 37% of the remainder, and so on.

Example 3.8

A 10 µF capacitor is charged to a potential of 20 V and then discharged through a 47 kΩ resistor. Determine the time taken for the capacitor voltage to fall below 10 V.

Table 3.1 Exponential growth and decay

t/(CR) or t/(L/R)	K[a] growth	decay
0.0	0.0000	1.0000
0.1	0.0951	0.9048
0.2	0.1812	0.8187 (1)
0.3	0.2591	0.7408
0.4	0.3296	0.6703
0.5	0.3935	0.6065
0.6	0.4511	0.5488
0.7	0.5034	0.4965
0.8	0.5506	0.4493
0.9	0.5934	0.4065
1.0	0.6321	0.3679
1.5	0.7769	0.2231
2.0	0.8647 (2)	0.1353
2.5	0.9179	0.0821
3.0	0.9502	0.0498
3.5	0.9698	0.0302
4.0	0.9817	0.0183
4.5	0.9889	0.0111
5.0	0.9933	0.0067

Notes: (1) See Example 3.9
(2) See Example 3.13
[a]K is the ratio of instantaneous value to final value (i.e. v_c/V_s etc.).

Figure 3.30   *C–R* integrating circuit

Figure 3.31   *C–R* differentiating circuit

Solution

The formula for exponential decay of voltage in the capacitor is:

$$v_c = V_s e^{-t/CR}$$

where $V_s = 20\,\text{V}$ and $CR = 10\,\mu\text{F} \times 47\,\text{k}\Omega = 0.47\,\text{s}$.

We need to find t when $v_c = 10\,\text{V}$.
Rearranging the formula to make t the subject gives:

$$t = -CR \times \ln(v_c/V_s)$$

thus

$$t = -0.47 \times \ln(10/20)$$

or

$$t = -470 \times -0.693 = 0.325\,\text{s}$$

In order to simplify the mathematics of exponential growth and decay, Table 3.1 provides an alternative tabular method that may be used to determine the voltage and current in a *C–R* circuit.

Example 3.9

A 150 µF capacitor is charged to a potential of 150 V. The capacitor is then removed from the charging source and connected to a 2 MΩ resistor. Determine the capacitor voltage 1 minute later.

Solution

We will solve this problem using Table 3.1 rather than the exponential formula.
First we need to find the time constant:

$$C \times R = 150\,\mu\text{F} \times 2\,\text{M}\Omega = 300\,\text{s}$$

Next we find the ratio of t to CR:
After 1 minute, $t = 60\,\text{s}$ therefore the ratio of t to CR is 60/300 or 0.2. Table 3.1 shows that when $t/CR = 0.2$, the ratio of instantaneous value to final value (K) is 0.8187.
Thus

$$v_c/V_s = 0.8187$$

or

$$v_c = 0.8187 \times V_s = 0.8187 \times 150\,\text{V} = 122.8\,\text{V}$$

Waveshaping with *C–R* networks

One of the most common applications of *C–R* networks is in waveshaping circuits. The circuits shown in Figs 3.30 and 3.31 function as simple square-to-triangle and square-to-pulse converters by, respectively, **integrating** and **differentiating** their inputs.

Figure 3.32 Waveforms for the integrating circuit (Fig. 3.30)

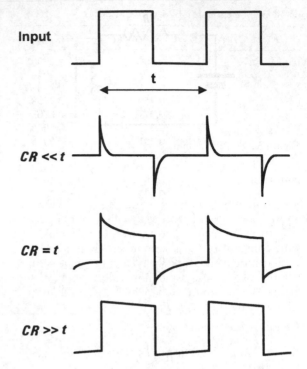

Figure 3.33 Waveforms for the differentiating circuit (Fig. 3.31)

The effectiveness of the simple **integrator circuit** shown in Fig. 3.30 depends very much upon the ratio of time constant ($C \times R$) to periodic time (t). The larger this ratio is, the more effective the circuit will be as an integrator. The effectiveness of the circuit of Fig. 3.30 is illustrated by the input and output waveforms shown in Fig. 3.32.

Similarly, the effectiveness of the simple **differentiator circuit** shown in Fig. 3.31 also depends very much upon the ratio of time constant ($C \times R$) to periodic time (t). The smaller this ratio is, the more effective the circuit will be as a differentiator. The effectiveness of the circuit of Fig. 3.31 is illustrated by the input and output waveforms shown in Fig. 3.33.

Example 3.10

A circuit is required to produce a train of alternating positive and negative pulses of short duration from a square wave of frequency 1 kHz. Devise a suitable C–R circuit and specify suitable values.

Solution

Here we require the services of a differentiating circuit along the lines of that shown in Fig. 3.31. In order that the circuit operates effectively as a differentiator, we need to make the time constant ($C \times R$) very much less than the periodic time of the input waveform (1 ms). Assuming that we choose a medium value for R of, say, 10 kΩ, the maximum value which we could allow C to have would be that which satisfies the equation:

$$C \times R = 0.1t$$

where $R = 10$ kΩ and $t = 1$ ms. Thus

$$C = \frac{0.1t}{R} = \frac{0.1 \times 1\,\text{ms}}{10\,\text{k}\Omega} = 0.1 \times 10^{-3} \times 10^{-4}$$
$$= 1 \times 10^{-8}$$

or

$$C = 10 \times 10^{-9} = 10\,\text{nF}$$

In practice, any value equal to or less than 10 nF would be adequate. A very small value (say below 1 nF) will, however, generate pulses of a very narrow width.

Example 3.11

A circuit is required to produce a triangular waveform from a square wave of frequency 1 kHz. Devise a suitable C–R arrangement and specify suitable values.

Figure 3.34   *L–R* circuit

Solution

This time we require an integrating circuit like that shown in Fig. 3.32. In order that the circuit operates effectively as an integrator, we need to make the time constant ($C \times R$) very much greater than the period time of the input waveform (1 ms). Assuming that we choose a medium value for R of, say, $10\,\text{k}\Omega$, the minimum value which we could allow C to have would be that which satisfies the equation:

$$C \times R = 10t$$

where $R = 10\,\text{k}\Omega$ and $t = 1\,\text{ms}$. Thus

$$C = \frac{10t}{R} = \frac{10 \times 1\,\text{ms}}{10\,\text{k}\Omega} = 10 \times 10^{-3} \times 10^{-4}$$

$$= 10 \times 10^{-7}$$

or

$$C = 1 \times 10^{-6} = 1\,\mu\text{F}$$

In practice, any value equal to or greater than $1\,\mu\text{F}$ would be adequate. A very large value (say above $10\,\mu\text{F}$) will, however, produce a triangle wave of very severely limited amplitude.

L–R circuits

Networks of inductors and resistors (known as *L–R* circuits) can also be used for timing and pulse shaping. In comparison with capacitors, however, inductors are somewhat more difficult to manufacture and are consequently more expensive. Inductors are also prone to losses and may also require screening to minimize the effects of stray magnetic coupling. Inductors are, therefore, generally unsuited to simple timing and waveshaping applications.

Figure 3.34 shows a simple *LR* network in which an inductor is connected to a constant voltage supply. When the supply is first connected, the current (i) will rise exponentially with time (as shown in Fig. 3.35). At the same time, the inductor voltage (V_L) will fall (as shown in Fig. 3.36). The rate of change of current with time will depend upon the ratio of inductance to resistance and is known as the time constant. Hence:

time constant, $t = L/R$

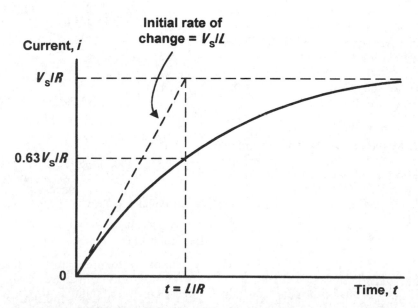

Figure 3.35   Exponential growth of current (*i*) in Fig. 3.34

Figure 3.36   Exponential decay of inductor voltage (v_L) in Fig. 3.34

where L is the value of inductance (H), R is the resistance (Ω), and t is the time constant (s).

The current flowing in the inductor (i) varies with time (t) according to the relationship:

$$i = V_s/R(1 - e^{-tR/L})$$

where V_s is the d.c. supply voltage, R is the resistance of the inductor, and L is the inductance.

The current (i) will initially be zero and will rise to approximately 63% of its maximum value (i.e. V_s/R) in a time interval equal to the time constant. At the end of the next interval of time equal to the time constant (i.e. after a total time of $2L/R$ has elapsed) the current will have risen by a further 63% of the remainder, and so on.

In theory, the current in the inductor will *never* become equal to V_s/R. However, after a period of time equal to $5L/R$, the current will to all intents and purposes be equal to V_s/L. At this point the current in the inductor will have risen to 99.3% of its final value.

The voltage developed across the inductor (V_L) varies with time (t) according to the relationship:

$$V_L = V_s e^{-tR/L}$$

where V_s is the d.c. supply voltage, R is the resistance of the inductor, and L is the inductance.

The inductor voltage will fall to approximately 37% of the initial voltage in a time equal to the time

constant. At the end of the next interval of time equal to the time constant (i.e. after a total time of $2L/R$ has elapsed) the voltage will have fallen by a further 37% of the remainder, and so on.

Example 3.12

A coil having inductance 6 H and resistance 24 Ω is connected to a 12 V d.c. supply. Determine the current in the inductor 0.1 s after the supply is first connected.

Solution

The formula for exponential growth of current in the coil is:

$$i = V_s/R(1 - e^{-tR/L})$$

where $V_s = 12$ V and $L/R = 6\,\text{H}/24\,\Omega = 0.25$ s.
We need to find i when $t = 0.1$ s

$$i = 12/24(1 - e^{-0.1/0.25}) = 0.5(1 - e^{-0.4})$$
$$= 0.5(1 - 0.67)$$

thus

$$i = 0.5 \times 0.33 = 0.165\,\text{A}.$$

In order to simplify the mathematics of exponential growth and decay, Table 3.1 provides an

Figure 3.37 *L–R integrating circuit*

Figure 3.38 *L–R differentiating circuit*

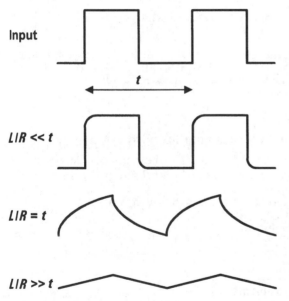

Figure 3.39 Waveforms for the integrating circuit (Fig. 3.37)

alternative tabular method that may be used to determine the voltage and current in an *L–R* circuit.

Example 3.13

A coil has an inductance of 100 mH and a resistance of 10 Ω. If the inductor is connected to a 5 V d.c. supply, determine the inductor voltage 20 ms after the supply is first connected.

Solution

We will solve this problem using Table 3.1 rather than the exponential formula.

First we need to find the time constant:

$$L/R = 0.1\text{H}/10\Omega = 0.01 \text{ s}.$$

Next we find the ratio of t to L/R:

when $t = 20$ ms the ratio of t to L/R is 0.02/0.01 or 2.

Table 3.1 shows that when $t/(L/R) = 2$, the ratio of instantaneous value to final value (K) is 0.8647.

Thus

$$v_\text{L}/V_\text{s} = 0.8647$$

or

$$v_\text{L} = 0.8647 \times v_\text{s} = 0.8647 \times 5 \text{ V} = 4.32 \text{ V}$$

Waveshaping with *L–R* networks

L–R networks are sometimes employed in waveshaping applications. The circuits shown in Figs 3.37 and 3.38 function as square-to-triangle and square-to-pulse converters by, respectively, integrating and differentiating their inputs.

The effectiveness of the simple **integrator circuit** shown in Fig. 3.37 depends very much upon the ratio of time constant (L/R) to periodic time (t). The larger this ratio is, the more effective the circuit will be as an integrator. The effectiveness of the circuit of Fig. 3.37 is illustrated by the input and output waveforms shown in Fig. 3.39.

Similarly, the effectiveness of the simple **differentiator** circuit shown in Fig. 3.38 also depends very much upon the ratio of time constant (L/R) to periodic time (t). The smaller this ratio is, the more effective the circuit will be as a differentiator. The effectiveness of the circuit of Fig. 3.38 is illustrated by the input and output waveforms shown in Fig. 3.40.

In practical waveshaping applications, *C–R* circuits are almost invariably superior to *L–R* circuits on the grounds of both cost and performance. Hence examples of the use of *LR* circuits in waveshaping applications have not been given.

Input

t

L/R << *t*

L/R = *t*

L/R >> *t*

Figure 3.40   Waveforms for the differentiating circuit (Fig. 3.38)

Formulae introduced in this chapter

Kirchhoff's current law:
(page 46)

Algebraic sum of currents = 0

Kirchhoff's voltage law:
(page 46)

Algebraic sum of e.m.f.s = algebraic sum of voltage drops

Potential divider:
(page 48)

$$V_{out} = V_{in} \times \frac{R_2}{R_1 + R_2}$$

Current divider:
(page 49)

$$I_{out} = I_{in} \times \frac{R_1}{R_1 + R_2}$$

Wheatstone bridge:
(page 50)

$R1/R2 = R3/R4$

$R_x = RV \times R2/R1$

Time constant of a *C–R* circuit:
(page 55)

$t = CR$

Capacitor voltage (charge):
(page 55)

$v_c = V_s(1 - e^{-t/CR})$

Capacitor current (charge):
(page 55)

$i = V_s e^{-t/CR}$

Capacitor voltage (discharge):
(page 55)

$v_c = V_s e^{-t/CR}$

Capacitor current (discharge):
(page 56)

$i = V_s e^{-t/CR}$

Time constant of an *L–R* circuit:
(page 59)

$t = L/R$

Inductor current (flux build up):
(page 60)

$i = V_s/R(1 - e^{-tR/L})$

Inductor voltage (flux build up):
(page 60)

$V_L = V_s e^{-tR/L}$

Problems

3.1   A power supply is rated at 500 mA maximum output current. If the supply delivers 150 mA to one circuit and 75 mA to another, how much current would be available for a third circuit?

3.2   A 15 V d.c. supply delivers a total current of 300 mA. If this current is shared equally between four circuits, determine the resistance of each circuit.

3.3   Determine the unknown current in each circuit shown in Fig. 3.42.

Circuit symbols introduced in this chapter

Figure 3.41

3.4 Determine the unknown voltage in each circuit shown in Fig. 3.43.

3.5 Determine all currents and voltages in Fig. 3.44.

Figure 3.42

3.6 Two resistors, one of $120\,\Omega$ and one of $680\,\Omega$, are connected as a potential divider across a 12 V supply. Determine the voltage developed across each resistor.

3.7 Two resistors, one of $15\,\Omega$ and one of $5\,\Omega$, are connected in parallel. If a current of 2 A is applied to the combination, determine the current flowing in each resistor.

3.8 A switched attenuator comprises five $1\,k\Omega$ resistors wired in series across a 5 V d.c. supply. If the output voltage is selected by means of a single-pole four-way switch, determine the voltage produced for each switch position.

3.9 A battery charger is designed to charge six 9 V batteries simultaneously from a raw 24 V d.c. supply. Each battery is connected to the 24 V supply via a separate resistor. If the batteries are to be charged at a nominal current of 10 mA each, determine the value of series resistance and the total drain on the 24 V supply.

(a)

(b)

Figure 3.43

(a)

(b)

Figure 3.44

Figure 3.45

Figure 3.46

Figure 3.47

Figure 3.48

3.10   A capacitor of 1 μF is charged from a 15 V d.c. supply via a 100 kΩ resistor. How long will it take for the capacitor voltage to reach 5 V?

3.11   A capacitor of 22 μF is charged to a voltage of 50 V. If the capacitor is then discharged using a resistor of 100 kΩ, determine the time taken for the capacitor voltage to reach 10 V.

3.12   An initially uncharged capacitor is charged from a 200 V d.c. supply through a 2 MΩ resistor. If it takes 50 s for the capacitor

voltage to reach 100 V, determine the value of capacitance.

3.13   An inductor has an inductance of 2.5 H and a resistance of 10 Ω. If the inductor is connected to a 5 V d.c. supply, determine the time taken for the current to grow to 200 mA.

3.14   Determine the Thévenin equivalent of the circuit shown in Fig. 3.45.

3.15   Determine the Norton equivalent of the circuit shown in Fig. 3.46.

3.16   The Thévenin equivalent of a network is shown in Fig. 3.47. Determine (a) the short-circuit output current and (b) the output voltage developed across a load of 200 Ω.

3.17   The Norton equivalent of a network is shown in Fig. 3.48. Determine (a) the open-circuit output voltage and (b) the output voltage developed across a load of 5 kΩ.

(Answers to these problems appear on page 260.)

4

Alternating voltage and current

This chapter introduces basic alternating current theory. We discuss the terminology used to describe alternating waveforms and the behaviour of resistors, capacitors, and inductors when an alternating current is applied to them. The chapter concludes by introducing another useful component, the transformer.

Signals can be conveyed using one or more of the properties of a waveform and sent using wires, cables, optical and radio links. Signals can also be processed in various ways using amplifiers, modulators, filters, etc. Signals are also classified as either **analogue** (continuously variable) or **digital** (based on discrete states).

a.c. versus d.c.

Direct currents are currents which, even though their magnitude may vary, essentially flow only in one direction. In other words, direct currents are unidirectional. Alternating currents, on the other hand, are bidirectional and continuously reverse their direction of flow. The polarity of the e.m.f. which produces an alternating current must consequently also be changing from positive to negative, and vice versa.

Alternating currents produce alternating potential differences (voltages) in the circuits in which they flow. Furthermore, in some circuits, alternating voltages may be superimposed on direct voltage levels (see Fig. 4.1). The resulting voltage may be unipolar (i.e. always positive or always negative) or bipolar (i.e. partly positive and partly negative).

Waveforms and signals

A graph showing the variation of voltage or current present in a circuit is known as a **waveform**. There are many common types of waveform encountered in electrical circuits including sine (or sinusoidal), square, triangle, ramp or sawtooth (which may be either positive or negative going), and pulse. **Complex waveforms** like speech or music usually comprise many components at different frequencies. **Pulse waveforms** are often categorized as either repetitive or non-repetitive (the former comprises a pattern of pulses which regularly repeats while the latter comprises pulses which constitute a unique event). Several of the most common waveform types are shown in Fig. 4.2.

Frequency

The frequency of a repetitive waveform is the number of cycles of the waveform which occur in unit time. Frequency is expressed in hertz (Hz). A frequency of 1 Hz is equivalent to one cycle per second. Hence, if a voltage has a frequency of 400 Hz, 400 cycles will occur in every second.

The equation for the voltage shown in Fig. 4.3 at a time, t, is:

$$v = V_{max} \sin(2\pi f t)$$

(a)

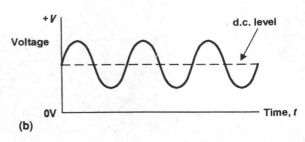

(b)

Figure 4.1 (a) Bipolar sine wave (this waveform swings symmetrically above and below 0 V); (b) unipolar sine wave (this waveform is superimposed on a d.c. level)

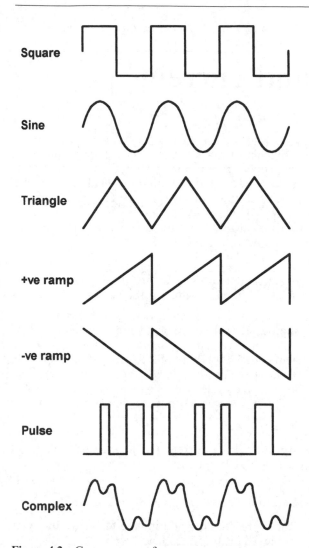

Square

Sine

Triangle

+ve ramp

-ve ramp

Pulse

Complex

Figure 4.2   Common waveforms

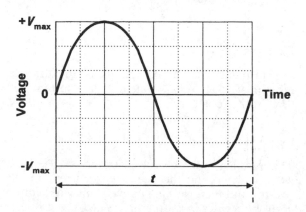

Figure 4.3   One cycle of a sine wave voltage showing its **periodic time**

where v is the instantaneous voltage, V_{max} is the maximum (or peak) voltage and f is the frequency.

Example 4.1

A sine wave voltage has a maximum value of 20 V and a frequency of 50 Hz. Determine the instantaneous voltage present (a) 2.5 ms and (b) 15 ms from the start of the cycle.

Solution

We can determine the voltage at any instant of time using:

$$v = V_{max} \sin(2\pi ft)$$

where $V_{max} = 20$ V and $f = 50$ Hz.
   In (a), $t = 2.5$ ms, hence:

$$v = 20 \sin(2\pi \times 50 \times 0.0025) = 20 \sin(0.785)$$
$$= 20 \times 0.707 = 14.14 \text{ V}$$

   In (b), $t = 15$ ms, hence:

$$v = 20 \sin(2\pi \times 50 \times 0.015) = 20 \sin(4.71)$$
$$= 20 \times -1 = -20 \text{ V}$$

Periodic time

The periodic time (or **period**) of a waveform is the time take for one complete cycle of the wave (see Fig. 4.3). The relationship between periodic time and frequency is thus:

$$t = 1/f \text{ or } f = 1/t$$

where t is the periodic time (in s) and f is the frequency (in Hz).

Example 4.2

A waveform has a frequency of 400 Hz. What is the periodic time of the waveform?

Solution

$$t = 1/f = 1/400 = 0.0025 \text{ s (or 2.5 ms)}$$

Hence the waveform has a periodic time of 2.5 ms.

Example 4.3

A waveform has a periodic time of 40 ms. What is its frequency?

Solution

$$f = 1/t = \frac{1}{40 \times 10^{-3}} = \frac{1}{0.04} = 25\,\text{Hz}$$

Average, peak, peak–peak, and r.m.s. values

The **average value** of an alternating current which swings symmetrically above and below zero will obviously be zero when measured over a long period of time. Hence average values of currents and voltages are invariably taken over one complete half-cycle (either positive or negative) rather than over one complete full-cycle (which would result in an average value of zero).

The **amplitude** (or **peak value**) of a waveform is a measure of the extent of its voltage or current excursion from the resting value (usually zero). The **peak-to-peak value** for a wave which is symmetrical about its resting value is twice its peak value (see Fig. 4.4).

The **r.m.s.** (or **effective**) **value** of an alternating voltage or current is the value which would produce the same heat energy in a resistor as a direct voltage or current of the same magnitude. Since the r.m.s. value of a waveform is very much dependent upon its shape, values are only meaningful when dealing with a waveform of known shape. Where the shape of a waveform is not specified, r.m.s. values are normally assumed to refer to sinusoidal conditions.

For a given waveform, a set of fixed relationships exist between average, peak, peak–peak, and r.m.s. values. The required multiplying factors are

Table 4.1 Multiplying factors for average, peak, peak–peak and r.m.s. values

Given quantity	Wanted quantity			
	Average	Peak	Peak–peak	r.m.s.
Average	1	1.57	3.14	1.11
Peak	0.636	1	2	0.707
Peak–peak	0.318	0.5	1	0.353
r.m.s.	0.9	1.414	2.828	1

summarized for sinusoidal voltages and currents in Table 4.1.

Example 4.4

A sinusoidal voltage has an r.m.s. value of 240 V. What is the peak value of the voltage?

Solution

The corresponding multiplying factor (found from Table 4.1) is 1.414. Hence:

$$V_{\text{pk}} = 1.414 \times V_{\text{r.m.s.}} = 1.414 \times 240 = 339.4\,\text{V}$$

Example 4.5

An alternating current has a peak–peak value of 50 mA. What is its r.m.s. value?

Solution

The corresponding multiplying factor (found from Table 4.1) is 0.353. Hence:

$$I_{\text{r.m.s.}} = 0.353 \times I_{\text{pk–pk}} = 0.353 \times 0.05 = 0.0177\,\text{A}$$
$$\text{(or } 17.7\,\text{mA)}$$

Example 4.6

A sinusoidal voltage 10 V pk–pk is applied to a resistor of 1 kΩ. What value of r.m.s. current will flow in the resistor?

Solution

This problem must be solved in two stages. First we will determine the peak–peak current in the resistor and then we shall convert this value into a corresponding r.m.s. quantity. Since

$$I = V/R,\ I_{\text{pk–pk}} = V_{\text{pk–pk}}/R$$

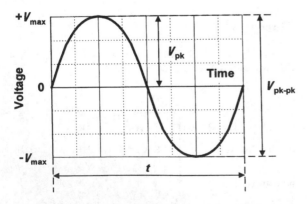

Figure 4.4 One cycle of a sine wave voltage showing its **peak** and **peak–peak** values

Hence:

$$I_{pk-pk} = 10\,V_{pk-pk}/1\,k\Omega = 10\,mA_{pk-pk}$$

The required multiplying factor (peak–peak to r.m.s.) is 0.353. Thus:

$$I_{r.m.s.} = 0.353 \times I_{pk-pk} = 0.353 \times 10\,mA = 3.53\,mA$$

Reactance

When alternating voltages are applied to capacitors or inductors the magnitude of the current flowing will depend upon the value of capacitance or inductance and on the frequency of the voltage. In effect, capacitors and inductors oppose the flow of current in much the same way as a resistor. The important difference being that the effective resistance (or reactance) of the component varies with frequency (unlike the case of a conventional resistor where the magnitude of the current does not change with frequency).

Capacitive reactance

The reactance of a capacitor is defined as the ratio of applied voltage to current and, like resistance, it is measured in ohms. The reactance of a capacitor is inversely proportional to both the value of capacitance and the frequency of the applied voltage. Capacitive reactance can be found by applying the following formula:

$$X_C = \frac{V_C}{I_C} = \frac{1}{2\pi f C}$$

where X_C is the reactance in ohms, f is the frequency in hertz, and C is the capacitance in Farads.

Capacitive reactance falls as frequency increases, as shown in Fig. 4.5.

The applied voltage, V_C, and current, I_C, flowing in a pure capacitive reactance will differ in phase by an angle of 90° or $\pi/2$ radians (the **current leads the voltage**). This relationship is illustrated in the current and voltage waveforms (drawn to a common time scale) shown in Fig. 4.6 and as a phasor diagram shown in Fig. 4.7.

Example 4.7

Determine the reactance of a 1 µF capacitor at (a) 100 Hz and (b) 10 kHz.

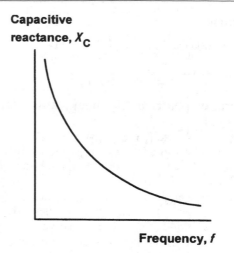

Figure 4.5 Variation of reactance with frequency for a capacitor

Figure 4.6 Voltage and current waveforms for a pure capacitor (the current leads the voltage by 90°)

Figure 4.7 Phasor diagram for a pure capacitor

Solution

(a) At 100 Hz

$$X_C = \frac{1}{2 \times \pi \times 100 \times 1 \times 10^{-6}}$$

or

$$X_C = \frac{0.159}{10^{-4}} = 0.159 \times 10^4$$

Thus

$$X_C = 1.59 \, k\Omega$$

(b) At 10 kHz

$$X_C = \frac{1}{2 \times \pi \times 10\,000 \times 1 \times 10^{-6}}$$

or

$$X_C = \frac{0.159}{10^{-2}} = 0.159 \times 10^2$$

Thus

$$X_C = 15.9 \, \Omega$$

Example 4.8

A 100 nF capacitor is to form part of a filter connected across a 240 V 50 Hz mains supply. What current will flow in the capacitor?

Solution

First we must find the reactance of the capacitor:

$$X_C = \frac{1}{2 \times \pi \times 50 \times 100 \times 10^{-9}} = 31.8 \, k\Omega$$

The r.m.s. current flowing in the capacitor will thus be:

$$I_C = \frac{V_C}{X_C} = \frac{240 \, V}{31.8 \, k\Omega} = 7.5 \, mA$$

Inductive reactance

The reactance of an inductor is defined as the ratio of applied voltage to current and, like resistance, it is measured in ohms. The reactance of an inductor is directly proportional to both the value of induct-

ance and the frequency of the applied voltage. Inductive reactance can be found by applying the formula:

$$X_L = \frac{V_L}{I_L} = 2\pi f L$$

where X_L is the reactance in Ω, f is the frequency in Hz, and L is the inductance in H.

Inductive reactance increases linearly with frequency as shown in Fig. 4.8.

The applied voltage current, I_L, and voltage, V_L, developed across a pure inductive reactance will differ in phase by an angle of 90° or $\pi/2$ radians (the **current lags the voltage**). This relationship is

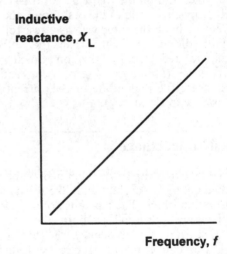

Figure 4.8   Variation of reactance with frequency for inductor

Figure 4.9   Voltage and current waveforms for a pure inductor (the voltage leads the current by 90°)

illustrated in the current and voltage waveforms (drawn to a common time scale) shown in Fig. 4.9 and as a phasor diagram shown in Fig. 4.10.

Example 4.9

Determine the reactance of a 10 mH inductor at (a) 100 Hz and (b) at 10 kHz.

Solution

(a) at 100 Hz

$$X_L = 2\pi \times 100 \times 10 \times 10^{-3}$$

Thus

$$X_L = 6.28 \, \Omega$$

(b) At 10 kHz

$$X_L = 2\pi \times 10\,000 \times 10 \times 10^{-3}$$

Thus

$$X_L = 628 \, \Omega$$

Example 4.10

A 100 mH inductor of negligible resistance is to form part of a filter which carries a current of 20 mA at 400 Hz. What voltage drop will be developed across the inductor?

Solution

The reactance of the inductor will be given by:

$$X_L = 2\pi \times 400 \times 100 \times 10^{-3} = 251 \, \Omega$$

The r.m.s. voltage developed across the inductor will be given by:

$$V_L = I_L \times X_L = 20 \, \text{mA} \times 251 \, \Omega = 5.02 \, \text{V}$$

In this example, it is important to note that we have assumed that the d.c. resistance of the inductor is negligible by comparison with its reactance. Where this is not the case, it is necessary to determine the impedance of the component and use this to determine the voltage drop.

Impedance

Figure 4.11 shows two circuits which contain both resistance and reactance. These circuits are said to exhibit impedance (a combination of resistance and reactance) which, like resistance and reactance, is measured in ohms. The impedance of the circuits shown in Fig. 4.11 is simply the ratio of supply voltage (V_S) to supply current (I_S).

The impedance of the simple $C-R$ and $L-R$ circuits shown in Fig. 4.11 can be found by using the impedance triangle shown in Fig. 4.12.

In either case, the impedance of the circuit is given by:

$$Z = \sqrt{(X^2 + R^2)}$$

and the phase angle (between V_S and I_S) is given by:

$$\phi = \tan^{-1}(X/R)$$

(a)

(b)

Figure 4.11 (a) C and R in series (this circuit exhibits impedance); (b) L and R in series (this circuit exhibits impedance)

Figure 4.10 Phasor diagram for a pure inductor

Figure 4.12 The impedance triangle

where Z is the impedance (in ohms), X is the reactance, either capacitive or inductive (expressed in ohms), R is the resistance (in ohms), and ϕ is the phase angle in radians.

Example 4.11

A $2\,\mu\text{F}$ capacitor is connected in series with a $100\,\Omega$ resistor across a $115\,\text{V}$ $400\,\text{Hz}$ a.c. supply. Determine the impedance of the circuit and the current taken from the supply.

Solution

First we must find the reactance of the capacitor, X_C:

$$X_C = \frac{1}{2\pi fC} = \frac{1}{6.28 \times 400 \times 2 \times 10^{-6}}$$
$$= \frac{10^6}{5024} = 199\,\Omega$$

Now we can find the impedance of the $C-R$ series circuit:

$$Z = \sqrt{(X_C^2 + R^2)} = \sqrt{(199^2 + 100^2)} = \sqrt{49\,601}$$
$$= 223\,\Omega$$

The current taken from the supply can now be found:

$$I_S = V_S/Z = 115/223 = 0.52\,\text{A}$$

Power factor

The power factor in an a.c. circuit containing resistance and reactance is simply the ratio of true power to apparent power. Hence:

$$\text{power factor} = \frac{\text{true power}}{\text{apparent power}}$$

The **true power** in an a.c. circuit is the power which is actually dissipated in the resistive component. Thus:

$$\text{true power} = I_S^2 \times R$$

The **apparent power** in an a.c. circuit is the power which is apparently consumed by the circuit and is the product of the supply current and supply voltage (note that this is not the same as the power which is actually dissipated as heat). Hence:

$$\text{apparent power} = I_S \times V_S\,\text{VA}$$

Hence

$$\text{power factor} = \frac{I_S^2 \times R}{I_S \times V_S} = \frac{I_S^2 \times R}{I_S \times (I_S \times Z)} = \frac{R}{Z}$$

From Fig. 4.12, $R/Z = \cos\phi$ thus:

$$\text{power factor} = R/Z = \cos\phi$$

Example 4.12

A choke having an inductance of $150\,\text{mH}$ and resistance of $250\,\Omega$ is connected to a $115\,\text{V}$ $400\,\text{Hz}$ a.c. supply. Determine the power factor of the choke and the current taken from the supply.

Solution

First we must find the reactance of the inductor, X_L.

$$X_L = 2\pi fL = 6.28 \times 400 \times 0.15 = 376.8\,\Omega$$

We can now determine the power factor:

$$\text{power factor} = R/Z = 250/376.8 = 0.663$$

The impedance of the choke (Z) will be given by:

$$Z = (X_L^2 + R^2) = (376.8^2 + 250^2) = 452\,\Omega$$

Finally, the current taken from the supply will be:

$$I_S = V_S/Z = 115/452 = 0.254\,\text{A}$$

L–C circuits

Two forms of $L-C$ circuits are illustrated in Figs 4.13 and 4.14. Figure 4.13 is a **series resonant** circuit while Fig. 4.14 constitutes a **parallel resonant**

Figure 4.13   Series resonant $L-C$ circuit

Figure 4.14   Parallel resonant $L-C$ circuit

circuit. The impedance of both circuits varies in a complex manner with frequency.

The impedance of the series circuit in Fig. 4.13 is given by:

$$Z = \sqrt{(X_L - X_C)^2}$$

where Z is the impedance of the circuit (in ohms), and X_L and X_C are the reactances of the inductor and capacitor respectively (both expressed in ohms).

The phase angle (between the supply voltage and current) will be $+\pi/2$ rad (i.e. $+90°$) when $X_L > X_C$ (above resonance) or $-\pi/2$ rad (or $-90°$) when $X_C > X_L$ (below resonance).

At a particular frequency (known as the **series resonant frequency**) the reactance of the capacitor (X_C) will be equal in magnitude (but of opposite sign) to that of the inductor (X_L). The impedance of the circuit will thus be zero at resonance. The supply current will have a maximum value at resonance (infinite in the case of a perfect series resonant circuit supplied from an ideal voltage source!).

The impedance of the parallel circuit in Fig. 4.14 is given by:

$$Z = \frac{X_L \times X_C}{\sqrt{(X_L - X_C)^2}}$$

where Z is the impedance of the circuit (in Ω), and X_L and X_C are the reactances of the inductor and capacitor, respectively (both expressed in Ω).

The phase angle (between the supply voltage and current) will be $+\pi/2$ rad (i.e. $+90°$) when $X_L > X_C$ (above resonance) or $-\pi/2$ rad (or $-90°$) when $X_C > X_L$ (below resonance).

At a particular frequency (known as the **parallel resonant frequency**) the reactance of the capacitor (X_C) will be equal in magnitude (but of opposite sign) to that of the inductor (X_L). At resonance, the denominator in the formula for impedance becomes zero and thus the circuit has an infinite impedance at resonance. The supply current will have a minimum value at resonance (zero in the case of a perfect parallel resonant circuit).

L–C–R networks

Two forms of $L-C-R$ network are illustrated in Figs 4.15 and 4.16; Fig. 4.15 is series resonant while Fig. 4.16 is parallel resonant. As in the case of their simpler $L-C$ counterparts, the impedance

Figure 4.15   Series resonant $L-C-R$ circuit

Figure 4.16   Parallel resonant $L-C-R$ circuit

of each circuit varies in a complex manner with frequency.

The impedance of the series circuit of Fig. 4.15 is given by:

$$Z = \sqrt{R^2 + (X_L - X_C)^2}$$

where Z is the impedance of the series circuit (in ohms), R is the resistance (in ohms), X_L is the inductive reactance (in ohms) and X_C is the capacitive reactance (also in ohms). At resonance the circuit has a minimum impedance (equal to R).

The phase angle (between the supply voltage and current) will be given by:

$$\phi = \tan^{-1} \frac{X_L - X_C}{R}$$

The impedance of the parallel circuit of Fig. 4.16 is given by:

$$Z = \frac{R \times X_L \times X_C}{\sqrt{(X_L^2 \times X_C^2) + R^2(X_L - X_C)^2}}$$

where Z is the impedance of the parallel circuit (in ohms), R is the resistance (in ohms), X_L is the inductive reactance (in ohms) and X_C is the capacitive reactance (also in ohms). At resonance the circuit has a maximum impedance (equal to R).

The phase angle (between the supply voltage and current) will be given by:

$$\phi = \tan^{-1} \frac{R(X_C - X_L)}{X_L \times X_C}$$

Resonance

The frequency at which the impedance is minimum for a series resonant circuit or maximum in the case of a parallel resonant circuit is known as the resonant frequency. The resonant frequency is given by:

$$f_0 = \frac{1}{2\pi\sqrt{LC}} \text{ Hz}$$

where f_0 is the resonant frequency (in hertz), L is the inductance (in henries) and C is the capacitance (in farads).

Typical impedance–frequency characteristics for series and parallel tuned circuits are shown in Figs

4.17 and 4.18. The series L–C–R tuned circuit has a minimum impedance at resonance (equal to R) and thus maximum current will flow. The circuit is consequently known as an **acceptor circuit**. The parallel L–C–R tuned circuit has a maximum impedance at resonance (equal to R) and thus minimum current will flow. The circuit is consequently known as a **rejector circuit**.

Quality factor

The quality of a resonant (or tuned) circuit is measured by its **Q-factor**. The higher the Q-factor, the sharper the response (narrower bandwidth), conversely the lower the Q-factor, the flatter the

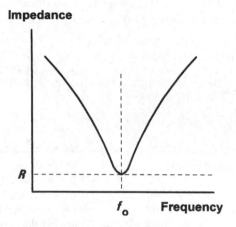

Figure 4.17 Impedance versus frequency for a series L–C–R **acceptor circuit**

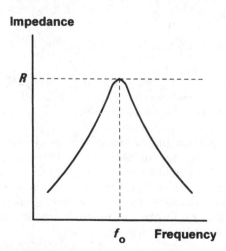

Figure 4.18 Impedance versus frequency for a parallel L–C–R **rejector circuit**

(a)

(b)

Figure 4.19 (a) Effect of Q-factor on the response of a parallel resonant circuit (the response is similar, but inverted, for a series resonant circuit); (b) bandwidth response (wider bandwidth), see Fig. 4.19. In the case of the series tuned circuit, the Q-factor will increase as the resistance, R, decreases. In the case of the parallel tuned circuit, the Q-factor will increase as the resistance, R, increases.

The response of a tuned circuit can be modified by incorporating a resistance of appropriate value either to 'dampen' or 'sharpen' the response.

The relationship between bandwidth and Q-factor is:

$$\text{bandwidth} = f_2 - f_1 = \frac{f_0}{Q} \text{ Hz}$$

Example 4.13

A parallel L–C circuit is to be resonant at a frequency of 400 Hz. If a 100 mH inductor is available, determine the value of capacitance required.

Solution

Re-arranging the formula $f_0 = 1/2\pi\sqrt{LC}$ to make C the subject gives:

$$C = \frac{1}{f_0^2(2\pi)^2 L}$$

Thus

$$C = \frac{1}{400^2 \times 39.4 \times 100 \times 10^{-3}} \text{ F}$$

or

$$C = \frac{1}{160 \times 10^3 \times 39.4 \times 100 \times 10^{-3}} \text{ F}$$

Hence $C = 1.58\,\mu\text{F}$.

This value can be realized from preferred values using a 2.2 µF capacitor connected in series with a 5.6 µF capacitor.

Example 4.14

A series L–C–R circuit comprises an inductor of 20 mH, a capacitor of 10 nF, and a resistor of 100 Ω. If the circuit is supplied with a sinusoidal signal of 1.5 V at a frequency of 2 kHz, determine the current supplied and the voltage developed across the resistor.

Solution

First we need to determine the values of inductive reactance (X_L) and capacitive reactance (X_C):

$$X_L = 2\pi f L = 6.28 \times 2 \times 10^3 \times 20 \times 10^{-3}$$

Thus $X_L = 251.2\,\Omega$.

$$X_C = \frac{1}{2\pi f C} = \frac{1}{6.28 \times 2 \times 10^3 \times 100 \times 10^{-9}}$$

Thus $X_C = 796.2\,\Omega$.

The impedance of the series circuit can now be calculated:

$$Z = \sqrt{R^2 + (X_L - X_C)^2}$$
$$= \sqrt{100^2 + (251.2 - 796.2)^2}$$

thus

$$Z = \sqrt{10\,000 + 297\,025} = \sqrt{307\,025} = 554\,\Omega$$

The current flowing in the series circuit will be given by:

$$I = V/Z = 1.5/554 = 2.7\,\text{mA}$$

The voltage developed across the resistor can now be calculated using:

$$V = IR = 2.7\,\text{mA} \times 100\,\Omega = 270\,\text{mV}$$

Transformers

Transformers provide us with a means of coupling a.c. power or signals from one circuit to another. Voltage may be **stepped-up** (secondary voltage greater than primary voltage) or **stepped-down** (secondary voltage less than primary voltage). Since no increase in power is possible (transformers are passive components like resistors, capacitors and inductors) an increase in secondary voltage can only be achieved at the expense of a corresponding reduction in secondary current, and vice versa (in fact, the secondary power will be very slightly less than the primary power due to losses within the transformer). Typical applications for transformers include stepping-up or stepping-down mains voltages in power supplies, coupling signals in AF amplifiers to achieve impedance matching and to isolate d.c. potentials associated with active components. The electrical characteristics of a transformer are determined by a number of factors including the core material and physical dimensions.

The specifications for a transformer usually include the rated primary and secondary voltages and currents the required power rating (i.e. the maximum power, usually expressed in volt–amperes, VA) which can be continuously delivered by the transformer under a given set of conditions), the frequency range for the component (usually stated as upper and lower working frequency limits), and the **regulation** of a transformer (usually expressed as a percentage of full-load). This last specification is a measure of the ability of a transformer to maintain its rated output voltage under load.

Table 4.2 summarizes the properties of three common types of transformer. Figure 4.20 shows the construction of a typical iron-cored power transformer.

Voltage and turns ratio

The principle of the transformer is illustrated in Fig. 4.21. The primary and secondary windings are wound on a common low-reluctance magnetic core. The alternating flux generated by the primary winding is therefore coupled into the secondary winding (very little flux escapes due to leakage). A sinusoidal current flowing in the primary winding produces a sinusoidal flux. At any instant the flux in the transformer is given by the equation:

$$\phi = \phi_{max} \sin(2\pi f t)$$

where ϕ_{max} is the maximum value of flux (in Webers), f is the frequency of the applied current (in hertz), and t is the time in seconds.

The r.m.s. value of the primary voltage (V_P) is given by:

$$V_P = 4.44 f N_P \phi_{max}$$

Similarly, the r.m.s. value of the secondary voltage (V_S) is given by:

$$V_S = 4.44 f N_S \phi_{max}$$

Figure 4.20   Construction of a typical iron-cored transformer

Now

$$V_P/V_S = N_P/N_S$$

where N_P/N_S is the **turns ratio** of the transformer.

Assuming that the transformer is loss-free, primary and secondary powers (P_P and P_S, respectively) will be identical. Hence:

$$P_P = P_S \text{ thus } V_P \times I_P = V_S \times I_S$$

Hence

$$V_P/V_S = I_S/I_P \text{ and } I_S/I_P = N_P/N_S$$

Finally, it is sometimes convenient to refer to a **turns-per-volt** rating for a transformer. This rating is given by:

$$\text{t.p.v.} = N_P/V_P = N_S/V_S$$

Table 4.2   Characteristics of common types of transformer

Parameter	Transformer type		
	Ferrite cored	Iron cored	Iron cored
Typical power rating	Less than 10 W	100 mW to 50 W	3 VA to 500 VA
Typical regulation	(see note)	(see note)	5% to 15%
Typical frequency range (Hz)	1 k to 10 M	50 to 20 k	45 to 400
Typical applications	Pulse circuits, RF power amplifiers	AF amplifiers	Power supplies

Note: Usually unimportant for this type of transformer.

Figure 4.21 The transformer principle

Example 4.15

A transformer has 2000 primary turns and 120 secondary turns. If the primary is connected to a 220 V r.m.s. a.c. mains supply, determine the secondary voltage.

Solution

Since $V_P/V_S = N_P/N_S$,

$$V_S = \frac{N_S \times V_P}{N_P} = \frac{120 \times 220}{2000} = 13.2 \text{ V}$$

Example 4.16

A transformer has 1200 primary turns and is designed to operate with a 200 V a.c. supply. If the transformer is required to produce an output of 10 V, determine the number of secondary turns required.

Solution

Since $V_P/V_S = N_P/N_S$,

$$N_S = \frac{N_P \times V_S}{V_P} = \frac{1200 \times 10}{200} = 60 \text{ turns}$$

Circuit symbols introduced in this chapter

Iron-cored transformer

Ferrite-cored transformer

Iron-cored transformer with multiple secondaries

Figure 4.22

Formulae introduced in this chapter

Sine wave voltage:
(page 66)

$v = V_{\max} \sin(2\pi ft)$

Frequency and periodic time:
(page 67)

$t = 1/f$ and $f = 1/t$

Peak and r.m.s. values for a sine wave:
(page 68)

$V_{pk} = 1.414 V_{r.m.s.}$
and $V_{r.m.s.} = 0.707 V_{pk}$

Inductive reactance:
(page 70)

$X_L = 2\pi fL$

Capacitive reactance:
(page 69)

$X_C = \dfrac{1}{2\pi fC}$

Impedance of C–R or L–R in series:
(page 71)

$$Z = \sqrt{(X^2 + R^2)}$$

Phase angle for C–R or L–R in series:
(page 71)

$$\phi = \tan^{-1}(X/R)$$

Power factor:
(page 72)

$$\text{P.F.} = \frac{\text{true power}}{\text{apparent power}}$$

$$\text{P.F.} = \cos\phi = R/Z$$

Resonant frequency of a series resonant tuned circuit:
(page 74)

$$f_0 = \frac{1}{2\pi\sqrt{LC}}$$

Bandwidth of a tuned circuit:
(page 75)

$$\text{B/W} = f_2 - f_1 = \frac{f_0}{Q}$$

Q-factor for a series tuned circuit:
(page 74)

$$Q = \frac{2\pi f L}{R}$$

Flux in a transformer:
(page 76)

$$\phi = \phi_{max}\sin(2\pi ft)$$

Transformer voltages:
(page 76)

$$V_P = 4.44 f N_P \phi_{max}$$
$$V_S = 4.44 f N_S \phi_{max}$$

Voltage and turns ratio:
(page 76)

$$V_P/V_S = N_P/N_S$$

Current and turns ratio:
(page 76)

$$I_S/I_P = N_P/N_S$$

Turns-per-volt:
(page 76)

$$N_P/V_P = N_S/V_S$$

Problems

4.1 A sine wave has a frequency of 250 Hz and an amplitude of 50 V. Determine its periodic time and r.m.s. value.

4.2 A sinusoidal voltage has an r.m.s. value of 240 V and a period of 16.7 ms. What is the frequency and peak value of the voltage?

4.3 Determine the frequency and peak–peak values of each of the waveforms shown in Fig. 4.23.

4.4 A sine wave has a frequency of 100 Hz and an amplitude of 20 V. Determine the instant-

Figure 4.23

aneous value of voltage (a) 2 ms and (b) 9 ms from the start of a cycle.

4.5 A sinusoidal current of 20 mA pk–pk flows in a resistor of 1.5 kΩ. Determine the r.m.s. voltage applied.

4.6 Determine the reactance of a 220 nF capacitor at (a) 20 Hz and (b) 5 kHz.

4.7 A 47 nF capacitor is connected across the 240 V 50 Hz mains supply. Determine the r.m.s. current flowing in the capacitor.

4.8 Determine the reactance of a 33 mH inductor at (a) 50 Hz and (b) 7 kHz.

4.9 A 10 mH inductor of negligible resistance is used to form part of a filter connected in series with a 50 Hz mains supply. What voltage drop will appear across the inductor when a current of 1.5 A is flowing?

4.10 A 10 μF capacitor is connected in series with a 500 Ω resistor across a 110 V 50 Hz a.c. supply. Determine the impedance of the circuit and the current taken from the supply.

4.11 A choke having an inductance of 1 H and resistance of 250 Ω is connected to a 220 V 60 Hz a.c. supply. Determine the power factor of the choke and the current taken from the supply.

4.12 A series-tuned *L–C* network is to be resonant at a frequency of 1.8 kHz. If a 60 mH inductor is available, determine the value of capacitance required.

4.13 A parallel resonant circuit employs a fixed inductor of 22 μH and a variable tuning capacitor. If the maximum and minimum values of capacitance are respectively 20 pF and 365 pF, determine the effective tuning range for the circuit.

4.14 A series *L–C–R* circuit comprises an inductor of 15 mH (with negligible resistance), a capacitor of 220 nF and a resistor of 100 Ω. If the circuit is supplied with a sinusoidal signal of 1.5 V at a frequency of 2 kHz, determine the current supplied and the voltage developed across the capacitor.

4.15 A 470 μH inductor has a resistance of 20 Ω. If the inductor is connected in series with a capacitor of 680 pF, determine the resonant frequency, *Q*-factor, and bandwidth of the circuit.

4.16 A transformer has 1600 primary turns and 120 secondary turns. If the primary is connected to a 240 V r.m.s. a.c. mains supply, determine the secondary voltage.

4.17 A transformer has 800 primary turns and 60 secondary turns. If the secondary is connected to a load resistance of 15 Ω, determine the value of primary voltage required to produce a power of 22.5 W in the load (assume that the transformer is loss-free).

(Answers to these problems appear on page 260.)

5

Semiconductors

This chapter introduces devices that are made from materials that are neither conductors nor insulators. These **semiconductor** materials form the basis of diodes, thyristors, triacs, transistors and integrated circuits. We start this chapter with a brief introduction to the principles of semiconductors before going on to examine the characteristics of each of the most common types of semiconductor.

In Chapter 1 we described the simplified structure of an atom and showed that it contains both negative charge carriers (electrons) and positive charge carriers (protons). Electrons each carry a single unit of negative electric charge while protons each exhibit a single unit of positive charge. Since atoms normally contain an equal number of electrons and protons, the net charge present will be zero. For example, if an atom has eleven electrons, it will also contain eleven protons. The end result is that the negative charge of the electrons will be exactly balanced by the positive charge of the protons.

Electrons are in constant motion as they orbit around the nucleus of the atom. Electron orbits are organized into shells. The maximum number of electrons present in the first shell is 2, in the second shell 8, and in the third, fourth and fifth shells it is 18, 32 and 50, respectively. In electronics, only the electron shell furthermost from the nucleus of an atom is important. It is important to note that the movement of electrons only involves those present in the outer **valence shell**.

If the valence shell contains the maximum number of electrons possible the electrons are rigidly bonded together and the material has the properties of an insulator. If, however, the valence shell does not have its full complement of electrons, the electrons can be easily loosened from their orbital bonds, and the material has the properties associated with an electrical conductor.

An isolated silicon atom contains four electrons in its valence shell. When silicon atoms combine to form a solid crystal, each atom positions itself between four other silicon atoms in such a way that the valence shells overlap from one atom to another.

This causes each individual valence electron to be shared by two atoms, as shown in Fig. 5.1. By sharing the electrons between four adjacent atoms, each individual silicon atom *appears* to have eight electrons in its valence shell. This sharing of valence electrons is called **covalent bonding**.

In its pure state, silicon is an insulator because the covalent bonding rigidly holds all of the electrons leaving no free (easily loosened) electrons to conduct current. If, however, an atom of a different element (i.e. an **impurity**) is introduced that has five electrons in its valence shell, a surplus electron will be present. These **free electrons** become available for use as **charge carriers** and they can be made to move through the lattice by applying an external potential difference to the material.

Similarly, if the impurity element introduced into the pure silicon lattice has three electrons in its valence shell, the absence of the fourth electron needed for proper covalent bonding will produce a

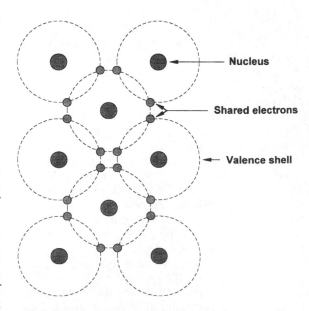

Figure 5.1 Lattice showing covalent bonding

number of spaces into which electrons can fit. These spaces are referred to as **holes**. Once again, current will flow when an external potential difference is applied to the material.

Regardless of whether the impurity element produces surplus electrons or holes, the material will no longer behave as an insulator, neither will it have the properties that we normally associate with a metallic conductor. Instead, we call the material a **semiconductor** – the term simply indicates that the substance is no longer a good insulator or a good conductor but is somewhere in between!

The process of introducing an atom of another (impurity) element into the lattice of an otherwise pure material is called **doping**. When the pure material is been doped with an impurity with five electrons in its valence shell (i.e. a **pentavalent impurity**) it will become an **N-type** material. If, however, the pure material is doped with an impurity having three electrons in its valence shell (i.e. a **trivalent impurity**) it will become **P-type** material. N-type semiconductor material contains an excess of negative charge carriers, and P-type material contains an excess of positive charge carriers.

Semiconductor diodes

When a junction is formed between N-type and P-type semiconductor materials, the resulting device is called a diode. This component offers an extremely low resistance to current flow in one direction and an extremely high resistance to current flow in the other. This characteristic allows the diode to be used in applications that require a circuit to behave differently according to the direction of current flowing in it.

An ideal diode would pass an infinite current in one direction and no current at all in the other direction. In addition, the diode would start to conduct current when the smallest of voltages was present. In practice, a small voltage must be applied before conduction takes place. Furthermore a small **leakage current** will flow in the **reverse direction**. This leakage current is usually a very small fraction of the current that flows in the **forward direction**.

If the P-type semiconductor material is made positive relative to the N-type material by an amount greater than its **forward threshold voltage** (about 0.6 V if the material is silicon and 0.2 V if the material is germanium), the diode will freely

Depletion layer where no free charge carriers exist

Figure 5.2 P–N junction diode

pass current. If, on the other hand, the P-type material is made negative relative to the N-type material, virtually no current will flow unless the applied voltage exceeds the maximum (breakdown) voltage that the device can withstand. Note that a normal diode will be destroyed if its **reverse breakdown voltage** is exceeded.

A semiconductor junction diode is shown in Fig. 5.2. The connection to the P-type material is referred to as the **anode** while that to the N-type material is called the **cathode**. With no externally applied potential, electrons from the N-type material will cross into the P-type region and fill some of the vacant holes. This action will result in the production of a region either side of the junction in which there are no free charge carriers. This zone is known as the **depletion region**.

Figure 5.3 shows a junction diode in which the anode is made positive with respect to the cathode. In this **forward-biased** condition, the diode freely passes current. Figure 5.4 shows a diode with the cathode made positive with respect to the cathode. In this **reverse-biased** condition, the diode passes a negligible amount of current. In the freely conducting forward-biased state, the diode acts rather like a closed switch. In the reverse-biased state, the diode acts like an open switch.

If a positive voltage is applied to the P-type material, the free positive charge carriers will be repelled and they will move away from the positive potential towards the junction. Likewise, the negative potential applied to the N-type material will cause the free negative charge carriers to move away from the negative potential towards the junction.

When the positive and negative charge carriers arrive at the junction, they will attract one another and combine (recall that unlike charges attract). As

Figure 5.3   Forward biased P–N junction

Figure 5.4   Reverse biased P–N junction

each negative and positive charge carrier combine at the junction, a new negative and positive charge carrier will be introduced to the semiconductor material from the voltage source. As these new charge carriers enter the semiconductor material, they will move toward the junction and combine. Thus, current flow is established and it will continue for as long as the voltage is applied.

As stated earlier, the **forward threshold voltage** must be exceeded before the diode will conduct. The forward threshold voltage must be high enough to completely remove the depletion layer and force charge carriers to move across the

junction. With silicon diodes, this forward threshold voltage is approximately 0.6 V to 0.7 V. With germanium diodes, the forward threshold voltage is approximately 0.2 V to 0.3 V.

Figure 5.5 shows typical characteristics for small germanium and silicon diodes. It is worth noting that diodes are limited by the amount of forward current and reverse voltage they can withstand. This limit is based on the physical size and construction of the diode.

In the case of a reverse biased diode, the P-type material is negatively biased relative to the N-type material. In this case, the negative potential

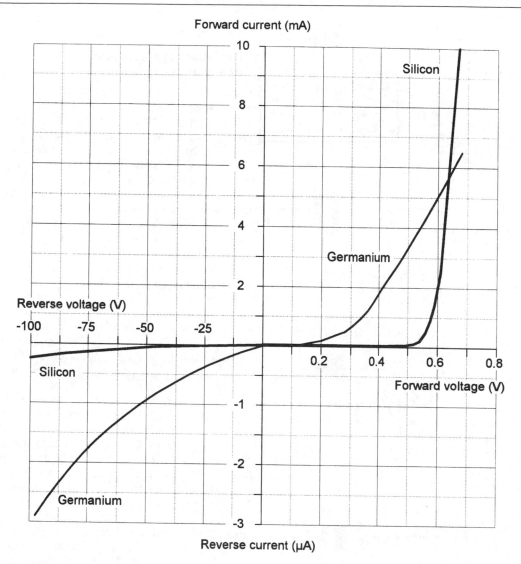

Figure 5.5 Typical diode characteristics

applied to the P-type material attracts the positive charge carriers, drawing them away from the junction. Likewise, the positive potential applied to the N-type material attracts the negative charge carriers away from the junction. This leaves the junction area depleted; virtually no charge carriers exist. Therefore, the junction area becomes an insulator, and current flow is inhibited.

The reverse bias potential may be increased to the reverse breakdown voltage for which the particular diode is rated. As in the case of the maximum forward current rating, the reverse breakdown voltage is specified by the manufac-

turer. The reverse breakdown voltage is usually very much higher than the forward threshold voltage. A typical general-purpose diode may be specified as having a forward threshold voltage of 0.6 V and a reverse breakdown voltage of 200 V. If the latter is exceeded, the diode may suffer irreversible damage. It is also worth noting that, where diodes are designed for use as rectifiers, manufacturers often quote **peak inverse voltage** (PIV) or **maximum reverse repetitive voltage** (V_{RRM}) rather than maximum reverse breakdown voltage.

Figure 5.6 shows a test circuit for obtaining diode characteristics (note that the diode must be

Figure 5.6   Diode test circuit

reverse connected in order to obtain the reverse characteristic).

Diode types

Diodes are often divided into **signal** or **rectifier** types according to their principal field of application. Signal diodes require consistent forward characteristics with low forward voltage drop. Rectifier diodes need to be able to cope with high values of reverse voltage and large values of forward current, consistency of characteristics is of secondary importance in such applications. Table 5.1 summarizes the characteristics of some common semiconductor diodes.

Example 5.1

The characteristic shown in Fig. 5.9 refers to a germanium diode. Determine the resistance of the diode when (a) $I_F = 2.5\,mA$ and (b) $V_F = 0.65\,V$.

Solution

See Fig. 5.9.

Zener diodes

Zener diodes are heavily doped silicon diodes which, unlike normal diodes, exhibit an abrupt reverse breakdown at relatively low voltages (typically less than 6 V). A similar effect occurs in less heavily doped diodes. These **avalanche diodes** also exhibit a rapid breakdown with negligible current flowing below the avalanche voltage and a relatively large current flowing once the avalanche voltage has been reached. For avalanche diodes, this breakdown voltage usually occurs at voltages above 6 V. In practice, however, both types of diode are referred to as zener diodes. A typical characteristic for a 5.1 V zener diode is shown in Fig. 5.7.

Whereas reverse breakdown is a highly undesirable effect in circuits that use conventional diodes, it can be extremely useful in the case of zener diodes where the breakdown voltage is precisely known. When a diode is undergoing reverse breakdown *and provided its maximum ratings are not exceeded* the voltage appearing across it will remain substantially constant (equal to the nominal zener voltage) regardless of the current flowing. This property makes the zener diode ideal for use as a **voltage regulator** (see Chapter 6).

Zener diodes are available in various families (according to their general characteristics, encapsulation and power ratings) with reverse breakdown (zener) voltages in the E12 and E24 series (ranging from 2.4 V to 91 V). Table 5.2 summarizes the characteristics of common zener diodes, while Fig. 5.10 shows typical encapsulations used for conventional diodes and zener diodes.

Figure 5.8 shows a test circuit for obtaining zener diode characteristics. The circuit is shown with the diode connected in the forward direction and it must be reverse connected in order to obtain the reverse characteristic.

Table 5.1   Characteristics of some common semiconductor diodes

Device	Material	PIV	I_F max.	I_R max.	Application
1N4148	Silicon	100 V	75 mA	25 nA	General purpose
1N914	Silicon	100 V	75 mA	25 nA	General purpose
AA113	Germanium	60 V	10 mA	200 µA	RF detector
OA47	Germanium	25 V	110 mA	100 µA	Signal detector
OA91	Germanium	115 V	50 mA	275 µA	General purpose
1N4001	Silicon	50 V	1 A	10 µA	Low-voltage rectifier
1N5404	Silicon	400 V	3 A	10 µA	High-voltage rectifier
BY127	Silicon	1250 V	1 A	10 µA	High-voltage rectifier

Figure 5.7 Typical characteristics for a 5.1 V zener diode

Table 5.2 Characteristics of some common zener diodes

BZY88 series	Miniature glass encapsulated diodes rated at 500 mW (at 25°C). Zener voltages range from 2.7 V to 15 V (voltages are quoted for 5 mA reverse current at 25°C)
BZX61 series	Encapsulated alloy junction rated at 1.3 W (25°C ambient). Zener voltages range from 7.5 V to 72 V
BZX85 series	Medium-power glass-encapsulated diodes rated at 1.3 W and offering zener voltages in the range 5.1 V to 62 V
BZY93 series	High-power diodes in stud mounting encapsulation. Rated at 20 W for ambient temperatures up to 75°C. Zener voltages range from 9.1 V to 75 V
1N5333 series	Plastic encapsulated diodes rated at 5 W. Zener voltages range from 3.3 V to 24 V

Diode coding

The European system for classifying semiconductor diodes involves an alphanumeric code which employs either two letters and three figures (general-purpose diodes) or three letters and two figures (special-purpose diodes). Table 5.3 shows how diodes are coded. Typical diode packages and markings (the stripe indicates the cathode connection) are shown in Fig. 5.10.

Figure 5.8   Zener diode test circuit

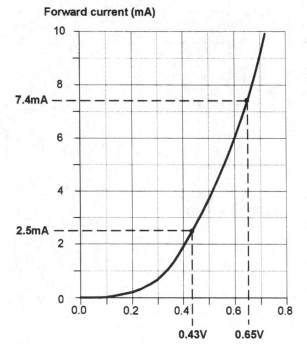

(a) R = 0.43V/2.5mA = 172 ohm
(b) R = 0.65V/7.4mA = 88 ohm

Figure 5.9

Example 5.2

Identify each of the following diodes:

(a) AA113
(b) BB105
(c) BZY88C4V7.

Solution

Diode (a) is a general-purpose germanium diode.
Diode (b) is a silicon variable capacitance diode.
Diode (c) is a silicon zener diode having ±5% tolerance and 4.7 V zener voltage.

Variable capacitance diodes

The capacitance of a reverse-biased diode junction will depend on the width of the depletion layer which, in turn, varies with the reverse voltage applied to the diode. This allows a diode to be used as a voltage controlled capacitor. Diodes that are specially manufactured to make use of this effect (and which produce comparatively large changes in capacitance for a small change in reverse voltage) are known as variable capacitance diodes (or **varicaps**). Such diodes are used (often in pairs) to provide tuning in radio and TV receivers. A typical characteristic for a variable capacitance diode is shown in Fig. 5.11. Table 5.4 summarizes the characteristics of several common variable capacitance diodes.

Thyristors

Thyristors (or **silicon controlled rectifiers**) are three-terminal devices which can be used for switching and a.c. power control.

Thyristors can switch very rapidly from a conducting to a non-conducting state. In the off state, the thyristor exhibits negligible leakage current, while in the on state the device exhibits very low resistance. This results in very little power loss within the thyristor even when appreciable power levels are being controlled. Once switched into the conducting state, the thyristor will remain conducting (i.e. it is latched in the on state) until the forward current is removed from the device. In d.c. applications this necessitates the interruption (or disconnection) of the supply before the device can be reset into its non-conducting state. Where the device is used with an alternating supply, the device will automatically become reset whenever the main supply reverses. The device can then be triggered on the next half-cycle having correct polarity to permit conduction.

Like their conventional silicon diode counterparts, thyristors have anode and cathode connections; control is applied by means of a gate

Table 5.3 Diode coding

First letter – semiconductor material: A Germanium
        B Silicon
        C Gallium arsenide, etc.
        D Photodiodes, etc.

Second letter – application: A General-purpose diode
        B Variable-capacitance diode
        E Tunnel diode
        P Photodiode
        Q Light emitting diode
        T Controlled rectifier
        X Varactor diode
        Y Power rectifier
        Z Zener diode

Third letter – in the case of diodes for specialized applications, the third letter does not generally have any particular significance

Zener diodes – zener diodes have an additional letter (which appears after the numbers) which denotes the tolerance of the zener voltage. The following letters are used:

        A ±1%
        B ±2%
        C ±5%
        D ±10%

Zener diodes also have additional characters which indicate the zener voltage (e.g. 9V1 denotes 9.1 V).

DO7   **DO14**

Figure 5.10 Some common diode packages

terminal (see Fig. 5.12). The device is triggered into the conducting (on state) by means of the application of a current pulse to this terminal. The effective triggering of a thyristor requires a **gate trigger** pulse having a fast rise time derived from a low-resistance source. Triggering can become erratic when insufficient gate current is available or when the gate current changes slowly. Table 5.5 summarizes the characteristics of several common thyristors.

Triacs

Triacs are a refinement of the thyristor which, when triggered, conduct on both positive and negative half-cycles of the applied voltage. Triacs have three terminals known as main-terminal one (MT1), main terminal two (MT2) and gate (G), as shown in Fig. 5.13. Triacs can be triggered by both positive and negative voltages applied between G and MT1 with positive and negative voltages present at MT2 respectively. Triacs thus provide **full-wave control** and offer superior performance in a.c. power control applications when compared with thyristors which only provide **half-wave control**. Table 5.6 summarizes the characteristics of several common triacs.

In order to simplify the design of triggering circuits, triacs are often used in conjunction with diacs (equivalent to a bi-directional zener diode). A typical **diac** conducts heavily when the applied voltage exceeds approximately 30 V in either direction. Once in the conducting state, the resistance of the diac falls to a very low value and thus a relatively large value of current will flow. The characteristic of a typical diac is shown in Fig. 5.14.

Figure 5.11   Typical characteristics for a variable capacitance diode

Table 5.4   Characteristics of some common variable capacitance diodes

Type	Capacitance	Capacitance ratio	Q-factor
1N5450	33 pF at 4 V	2.6 for 4 V to 60 V	350
MV1404	50 pF at 4 V	>10 for 2 V to 10 V change	200
MV2103	10 pF at 4 V	2 for 4 V to 60 V	400
MV2115	100 pF at 4 V	2.6 for 4 V to 60 V	100

Table 5.5   Characteristics of some common thyristors

Type	$I_{F(AV)}$ (A)	V_{RRM} (V)	V_{GT} (V)	I_{GT} (mA)
2N4444	5.1	600	1.5	30
BT106	1	700	3.5	50
BT152	13	600	1	32
BTY79–400R	6.4	400	3	30
TIC106D	3.2	400	1.2	0.200
TIC126D	7.5	400	2.5	20

Figure 5.12   Thyristor connections

Light emitting diodes

Light emitting diodes (LED) can be used as general-purpose indicators and, compared with conventional filament lamps, operate from significantly smaller voltages and currents. LEDs are also very much more reliable than filament lamps. Most LEDs will provide a reasonable level of light output when a forward current of between 5 mA and 20 mA is applied.

Light emitting diodes are available in various formats with the round types being most popular. Round LEDs are commonly available in the 3 mm and 5 mm (0.2 inch) diameter plastic packages and also in a 5 mm × 2 mm rectangular format. The viewing angle for round LEDs tends to be in the region of 20° to 40°, whereas for rectangular types this is increased to around 100°. Table 5.7 summarizes the characteristics of several common types of LED.

In order to limit the forward current of an LED to an appropriate value, it is usually necessary to include a fixed resistor in series with an LED indicator, as shown in Fig. 5.15. The value of the resistor may be calculated from:

$$R = \frac{V - V_F}{I}$$

Main terminal 2 (mt2)

Gate (g)

Main terminal 1 (mt1)

Figure 5.13   Triac connections

Table 5.6   Characteristics of some common triacs

Type	$I_{T(RMS)}$ (A)	V_{RRM} (V)	V_{GT} (V)	$I_{GT(TYP)}$ (mA)
2N6075	4	600	2.5	5
BT139	15	600	1.5	5
TIC206M	4	600	2	5
TIC226M	8	600	2	50

where V_F is the forward voltage drop produced by the LED and V is the applied voltage. Note that it is usually safe to assume that V will be 2 V and choose the nearest preferred value for R.

Example 5.3

An LED is to be used to indicate the presence of a 21 V d.c. supply rail. If the LED has a nominal forward voltage of 2.2 V, and is rated at a current of 15 mA, determine the value of series resistor required.

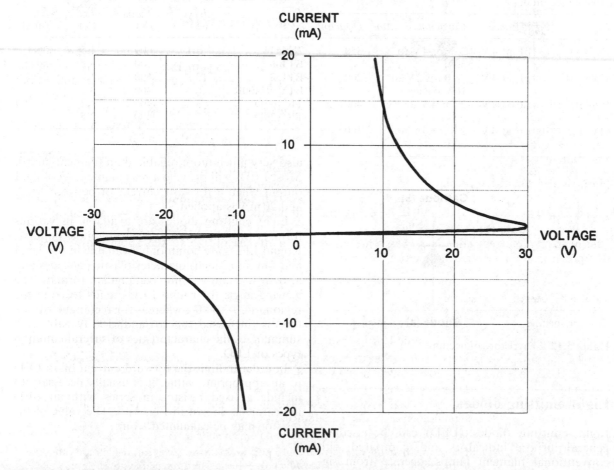

Figure 5.14   Diac characteristics

Table 5.7 Characteristics of some common types of LED

Parameter	Types of LED			
	Standard	Standard	High efficiency	High intensity
Diameter (mm)	3	5	5	5
Max. forward current (mA)	40	30	30	30
Typical forward current (mA)	12	10	7	10
Typical forward voltage drop (V)	2.1	2.0	1.8	2.2
Max. reverse voltage (V)	5	3	5	5
Max. power dissipation (mW)	150	100	27	135
Peak wavelength (nm)	690	635	635	635

Figure 5.15 Use of a current limiting resistor with an LED

Solution

Here we can use the formula:

$$R = \frac{V - V_F}{I} = \frac{21\,\text{V} - 2.2\,\text{V}}{15\,\text{mA}} = \frac{18.8\,\text{V}}{15\,\text{mA}} = 1.25\,\text{k}\Omega$$

The nearest preferred value is $1.2\,\text{k}\Omega$. The power dissipated in the resistor will be given by:

$$P = I \times V = 15\,\text{mA} \times 18.8\,\text{V} = 280\,\text{mW}$$

Hence the resistor should be rated at $0.33\,\text{W}$, or greater.

Bipolar transistors

Transistor is short for **transfer resistor**, a term which provides something of a clue as to how the device operates; the current flowing in the output circuit is determined by the current flowing in the input circuit. Since transistors are three-terminal devices, one electrode must remain common to both the input and the output.

Transistors fall into two main categories (bipolar and field-effect) and are also classified according to the semiconductor material employed (silicon or germanium) and to their field of application (e.g. general purpose, switching, high-frequency, etc.). Various classes of transistor are available according to the application concerned (see Table 5.8).

Table 5.8 Classes of transistor

Low-frequency	Transistors designed specifically for audio frequency applications (below 100 kHz)
High-frequency	Transistors designed specifically for radio frequency applications (100 kHz and above)
Power	Transistors that operate at significant power levels (such devices are often sub-divided into audio frequency and radio frequency power types)
Switching	Transistors designed for switching applications
Low-noise	Transistors that have low-noise characteristics and which are intended primarily for the amplification of low-amplitude signals
High-voltage	Transistors designed specifically to handle high voltages
Driver	Transistors that operate at medium power and voltage levels and which are often used to precede a final (power) stage which operates at an appreciable power level

Transistor coding

The European system for classifying transistors involves an alphanumeric code which employs either two letters and three figures (general-purpose transistors) or three letters and two figures (special-purpose transistors). Table 5.9 shows how transistors are coded.

Example 5.4

Identify each of the following transistors:

(a) AF115
(b) BC109
(c) BD135
(d) BFY51.

Solution

Transistor (a) is a general-purpose, low-power, high-frequency germanium transistor.

Transistor (b) is a general-purpose, low-power, low-frequency silicon transistor.

Transistor (c) is a general-purpose, high-power, low-frequency silicon transistor.

Transistor (d) is a special-purpose, low-power, high-frequency silicon transistor.

Transistor operation

Bipolar transistors generally comprise NPN or PNP junctions of either **silicon** (Si) or **germanium** (Ge) material (see Figs 5.16 and 5.17). The junctions are, in fact, produced in a single slice of silicon by diffusing impurities through a photographically reduced mask. Silicon transistors are superior when compared with germanium transistors in the vast majority of applications (particularly at high temperatures) and thus germanium devices are very rarely encountered.

Figures 5.18 and 5.19, respectively, show a simplified representation of NPN and PNP tran-

Table 5.9 Transistor coding

First letter – semiconductor material:	A	Germanium
	B	Silicon
Second letter – application:	C	Low-power, low-frequency
	D	High-power, low-frequency
	F	Low-power, high-frequency
	L	High-power, high-frequency

Third letter – in the case of transistors for specialized and industrial applications, the third letter does not generally have any particular significance

Figure 5.16 NPN transistor construction

Figure 5.17   PNP transistor construction

Figure 5.18   Simplified model of an NPN transistor

Figure 5.19   Simplified model of a PNP transistor

sistors together with their circuit symbols. In either case the electrodes are labelled **collector, base** and **emitter**. Note that each junction within the transistor, whether it be collector–base or base–emitter, constitutes a P–N junction.

Figures 5.20 and 5.21, respectively, show the normal bias voltages applied to NPN and PNP transistors. Note that the base–emitter junction is forward biased and the collector–base junction is reverse biased. The base region is, however, made very narrow so that carriers are swept across it from emitter to collector and only a relatively small current flows in the base. To put this into context, the current flowing in the emitter circuit is typically 100 times greater than that flowing in the

Figure 5.20   Bias voltages and current flow in an NPN transistor

Figure 5.21   Bias voltages and current flow in a PNP transistor

Figure 5.22   Typical input characteristic for a small-signal NPN transistor operating in common-emitter mode

base. The direction of conventional current flow is from emitter to collector in the case of a PNP transistor, and collector to emitter in the case of an NPN device.

The equation which relates current flow in the collector, base, and emitter currents is:

$$I_E = I_B + I_C$$

where I_E is the emitter current, I_B is the base current, and I_C is the collector current (all expressed in the same units).

Bipolar transistor characteristics

The characteristics of a transistor are often presented in the form of a set of graphs relating voltage and current present at the transistor's terminals.

A typical **input characteristic** (I_B plotted against V_{BE}) for a small-signal general-purpose NPN transistor operating in common-emitter mode

(see Chapter 7) is shown in Fig. 5.22. This characteristic shows that very little base current flows until the base–emitter voltage (V_{BE}) exceeds 0.6 V. Thereafter, the base current increases rapidly (this characteristic bears a close resemblance to the forward part of the characteristic for a silicon diode, see Fig. 5.5).

Figure 5.23 shows a typical **output characteristic** (I_C plotted against V_{CE}) for a small-signal general-purpose NPN transistor operating in common-emitter mode (see Chapter 7). This characteristic comprises a family of curves, each relating to a different value of base current (I_B). It is worth taking a little time to get familiar with this characteristic as we shall be putting it to good use in Chapter 7. In particular it is important to note the 'knee' that occurs at values of V_{CE} of about 2 V. Also, note how the curves become flattened above this value with the collector current (I_C) not changing very greatly for a comparatively large change in collector–emitter voltage (V_{CE}).

Finally, a typical **transfer characteristic** (I_C plotted against I_B) for a small-signal general-purpose NPN transistor operating in common-emitter mode (see Chapter 7) is shown in Fig. 5.24. This characteristic shows an almost linear relationship between collector current and base current (i.e. doubling the value of base current produces double the value of collector current, and so on). This characteristic is reasonably independent of the value of collector–emitter voltage (V_{CE}) and thus only a single curve is used.

Current gain

The current gain offered by a transistor is a measure of its effectiveness as an amplifying device. The most commonly quoted parameter is that which relates to **common-emitter mode**. In this mode, the input current is applied to the base and the output current appears in the collector (the emitter is effectively common to both the input and output circuits).

The common-emitter current gain is given by:

$$h_{FE} = I_C/I_B$$

where h_{FE} is the **hybrid parameter** which represents **large signal** (d.c.) **forward current gain**, I_C is the collector current, and I_B is the base current. When small (rather than large) signal operation is considered, the values of I_C and I_B are incremental

Figure 5.23   Typical family of output characteristics for a small-signal NPN transistor operating in common-emitter mode

(i.e. small changes rather than static values). The current gain is then given by:

$$h_{fe} = I_c/I_b$$

where h_{fe} is the **hybrid parameter** which represents **small signal** (a.c.) **forward current gain**, I_c is the change in collector current which results from a corresponding change in base current, I_b.

Values of h_{FE} and h_{fe} can be obtained from the transfer characteristic (I_C plotted against I_B) as shown in Figs 5.25 and 5.26. Note that h_{FE} is found from corresponding static values while h_{fe} is found by measuring the slope of the graph. Also note that, if the transfer characteristic is linear, there is little (if any) difference between h_{FE} and h_{fe}.

It is worth noting that current gain (h_{fe}) varies with collector current. For most small-signal transistors, h_{fe} is a maximum at a collector current in the range 1 mA and 10 mA. Current gain also falls to very low values for power transistors when

operating at very high values of collector current. Furthermore, most transistor parameters (particularly common-emitter current gain, h_{fe}) are liable to wide variation from one device to the next. It is, therefore, important to design circuits on the basis of the minimum value for h_{fe} in order to ensure successful operation with a variety of different devices.

Transistor parameters are listed in Table 5.10, while Table 5.11 shows the characteristics of several common types of bipolar transistor. Finally, Fig. 5.27 shows a test circuit for obtaining NPN transistor characteristics (the arrangement for a PNP transistor is similar but all meters and supplies must be reversed).

Example 5.5

A transistor operates with $I_C = 30\,\text{mA}$ and $I_B = 600\,\mu\text{A}$. Determine the value of I_E and h_{FE}.

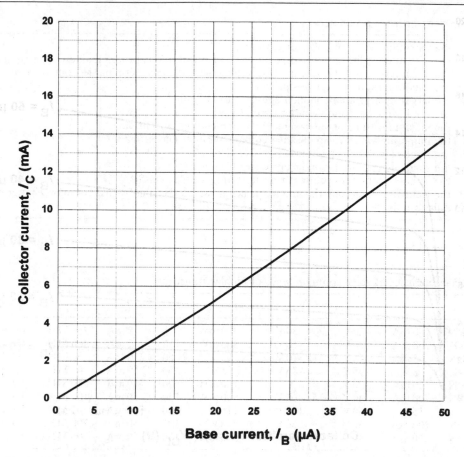

Figure 5.24 Typical transfer characteristic for a small-signal NPN transistor operating in common-emitter mode

Figure 5.25 Determining the static value of current gain (h_{FE}) from the transfer characteristic

Figure 5.26 Determining the small-signal value of current gain (h_{fe}) from the transfer characteristic

Table 5.10   Bipolar transistor parameters

I_C max.	– the maximum value of collector current
V_{CEO} max.	– the maximum value of collector–emitter voltage with the base terminal left open-circuit
V_{CBO} max.	– the maximum value of collector–base voltage with the base terminal left open-circuit
P_t max.	– the maximum total power dissipation
h_{FE}	– the large signal (static) common-emitter current gain
h_{fe}	– the small-signal common-emitter current gain
h_{fe} max.	– the maximum value of small-signal common-emitter current gain
h_{fe} min.	– the minimum value of small-signal common-emitter current gain
h_{ie}	– the small-signal input resistance (see Chapter 7)
h_{oe}	– the small-signal output conductance (see Chapter 7)
h_{re}	– the small-signal reverse current transfer ratio (see Chapter 7)
f_t typ.	– the transition frequency (i.e. the frequency at which the small-signal common-emitter current gain falls to unity

Table 5.11   Characteristics of some common types of bipolar transistor

Device	Type	I_c max.	V_{ceo} max.	V_{cbo} max.	P_t max.	h_{fe}	at I_c	f_t typ.	Application
BC108	NPN	100 mA	20 V	30 V	300 mW	125	2 mA	250 MHz	General purpose
BCY70	PNP	200 mA	−40 V	−50 V	360 mW	150	2 mA	200 MHz	General purpose
BD131	NPN	3 A	45 V	70 V	15 W	50	250 mA	60 MHz	AF power
BD132	PNP	3 A	−45 V	−45 V	15 W	50	250 mA	60 MHz	AF power
BF180	NPN	20 mA	20 V	20 V	150 mW	100	10 mA	650 MHz	RF amplifier
2N3053	NPN	700 mA	40 V	60 V	800 mW	150	50 mA	100 MHz	Driver
2N3055	NPN	15 A	60 V	100 V	115 W	50	500 mA	1 MHz	LF power
2N3866	NPN	400 mA	30 V	30 V	3 W	105	50 mA	700 MHz	RF driver
2N3904	NPN	200 mA	40 V	60 V	310 mW	150	50 mA	300 MHz	Switching

Solution

The value of I_E can be calculated from $I_E = I_C + I_B$, thus $I_E = 30 + 0.6 = 30.6$ mA.

The value of h_{FE} can be calculated from $h_{FE} = I_C/I_B = 30/0.6 = 50$.

Example 5.6

A transistor operates with a collector current of 97 mA and an emitter current of 98 mA. Determine the value of base current and common-emitter current gain.

Solution

Since $I_E = I_B + I_C$, the base current will be given by:

$$I_B = I_E - I_C = 98 - 97 = 1 \text{ mA}$$

The common-emitter current gain (h_{FE}) will be given by:

$$h_{FE} = I_C/I_B = 97/1 = 97$$

Example 5.7

An NPN transistor is to be used in a regulator circuit in which a collector current of 1.5 A is to be controlled by a base current of 50 mA. What value of h_{FE} will be required? If the device is to be operated with $V_{CE} = 6$ V, which transistor selected from Table 5.11 would be appropriate for this application and why?

Solution

The required current gain can be found from:

$$h_{FE} = I_C/I_B = 1.5 \text{ A}/50 \text{ mA} = 1500 \text{ mA}/50 \text{ mA}$$
$$= 30$$

Figure 5.27 NPN transistor test circuit (the arrangement for a PNP transistor is similar but all meters and supplies must be reversed)

The most appropriate device would be the BD131. The only other device capable of operating at a collector current of 1.5 A would be a 2N3055. The collector power dissipation will be given by:

$$P_C = I_C \times V_{CE} = 1.5\,A \times 6\,V = 9\,W$$

However, the 2N3055 is rated at 115 W maximum total power dissipation and this is more than ten times the power required.

Example 5.8

A transistor is used in a linear amplifier arrangement. The transistor has small and large signal current gains of 200 and 175, respectively, and bias is arranged so that the static value of collector current is 10 mA. Determine the value of base bias current and the change of output (collector) current that would result from a 10 μA change in input (base) current.

Solution

The value of base bias current can be determined from:

$$I_B = I_C/h_{FE} = 10\,mA/200 = 50\,\mu A$$

The change of collector current resulting from a 10 μA change in input current will be given by:

$$I_c = h_{fe} \times I_b = 175 \times 10\,\mu A = 1.75\,mA$$

Field effect transistors

Field effect transistors (FET) comprise a channel of P-type or N-type material surrounded by material of the opposite polarity. The ends of the channel (in which conduction takes place) form electrodes known as the **source** and **drain**. The effective width of the channel (in which conduction takes place) is controlled by a charge placed on the third (**gate**) electrode. The effective resistance between the source and drain is thus determined by the voltage present at the gate.

Field effect transistors are available in two basic forms; **junction gate** and **insulated gate**. The gate–source junction of a junction gate field effect transistor (JFET) is effectively a reverse-biased P–N junction. The gate connection of an insulated gate field effect transistor (IGFET), on the other hand, is insulated from the channel and charge is capacitively coupled to the channel. To keep things simple, we will consider only JFET devices in this book. Figure 5.28 shows the basic construction of an N-channel JFET.

JFETs offer a very much higher input resistance when compared with bipolar transistors. For example, the input resistance of a bipolar transistor operating in common-emitter mode (see

Figure 5.28 N-channel junction gate JFET construction

Chapter 7) is usually around 2.5 kΩ. A JFET transistor operating in equivalent common-source mode (see Chapter 7) would typically exhibit an input resistance of 100 MΩ! This feature makes JFET devices ideal for use in applications where a very high input resistance is desirable.

FET characteristics

As with bipolar transistors, the characteristics of a FET are often presented in the form of a set of graphs relating voltage and current present at the transistor's terminals.

A typical **mutual characteristic** (I_D plotted against V_{GS}) for a small-signal general-purpose N-channel JFET operating in common-source mode (see Chapter 7) is shown in Fig. 5.29. This characteristic shows that the drain current is progressively reduced as the gate–source voltage is made more negative. At a certain value of V_{GS} the drain current falls to zero and the device is said to be **cut-off**.

Figure 5.30 shows a typical **output characteristic** (I_D plotted against V_{GS}) for a small-signal general-purpose N-channel JFET operating in common-source mode (see Chapter 7). This characteristic comprises a family of curves, each relating to a different value of gate–source voltage (V_{GS}). Once again, it is worth taking a little time to get familiar with this characteristic as we shall be using it again in Chapter 7 (you might also like to compare this characteristic with the output characteristic for a transistor operating in common-emitter mode – see Fig. 5.23).

Once again, the characteristic curves have a 'knee' that occurs at low values of V_{DS}. Also, note how the curves become flattened above this value with the drain current (I_D) not changing very greatly for a comparatively large change in drain-source voltage (V_{DS}). These character-istics are, in fact, even flatter than those for a bipolar transistor. Because of their flatness, they are often said to represent a **constant current** characteristic.

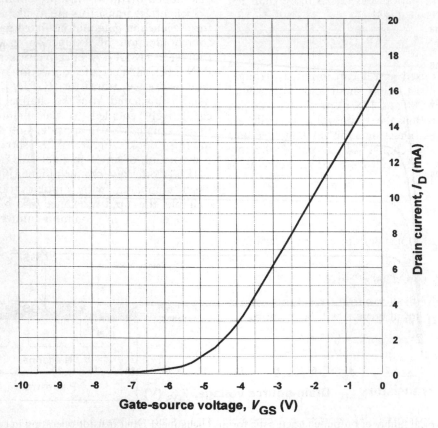

Figure 5.29 Typical mutual characteristic for an N-channel FET transistor operating in common-source mode

FET parameters

The gain offered by a field effect transistor is normally expressed in terms of its **forward transfer conductance** (g_{fs} or Y_{fs}) in **common source** mode. In this mode, the input voltage is applied to the gate and the output current appears in the drain (the source is effectively common to both the input and output circuits).

The common source forward transfer conductance is given by:

$$g_{fs} = I_d/V_{gs}$$

where I_d is the change in drain current resulting from a corresponding change in gate–source voltage (V_{gs}). The units of forward transfer conductance are siemens (S).

Forward transfer conductance (g_{fs}) varies with drain current collector current. For most small-signal devices, g_{fs} is quoted for values of drain current between 1 mA and 10 mA.

Most FET parameters (particularly forward transfer conductance) are liable to wide variation from one device to the next. It is, therefore, important to design circuits on the basis of the minimum value for g_{fs} in order to ensure successful operation with a variety of different devices.

The characteristics of several common N-channel field effect transistors are shown in Table 5.13.

Figure 5.31 shows a test circuit for obtaining the characteristics of an N-channel FET (the arrangement for a P-channel FET is similar but all meters and supplies must be reversed).

Example 5.9

A FET operates with a drain current of 100 mA and a gate–source bias of −1 V. If the device has a g_{fs} of 0.25 S, determine the change in drain current if the bias voltage increases to −1.1 V.

Figure 5.30 Typical family of output characteristic for an N-channel FET transistor operating in common-source mode

Table 5.12   FET parameters

I_D max.	– the maximum drain current
V_{DS} max.	– the maximum drain–source voltage
V_{GS} max.	– the maximum gate–source voltage
P_D max.	– the maximum drain power dissipation
t_r typ.	– the typical output rise-time in response to a perfect rectangular pulse input
t_f typ.	– the typical output fall-time in response to a perfect rectangular pulse input
$R_{DS(on)}$ max.	– the maximum value of resistance between drain and source when the transistor is in the conducting (on) state

Solution

The change in gate–source voltage (V_{gs}) is −0.1 V and the resulting change in drain current can be determined from:

$$I_d = g_{fs} \times V_{gs} = 0.25\,\text{S} \times -0.1\,\text{V} = -0.025\,\text{A}$$
$$= -25\,\text{mA}$$

The new value of drain current will thus be (100 mA − 25 mA) or 75 mA.

Transistor packages

A wide variety of packaging styles are used for transistors. Small-signal transistors tend to have either plastic packages (e.g. TO92) or miniature metal cases (e.g. TO5 or TO18). Medium and high-power devices may also be supplied in plastic cases but these are normally fitted with integral metal heat-sinking tabs (e.g. TO126, TO218 or TO220) in order to conduct heat away from the junction. Some older power transistors are supplied in metal cases (either TO66 or TO3). Several popular transistor case styles are shown in Fig. 5.32.

Integrated circuits

Integrated circuits are complex circuits fabricated on a small slice of silicon. Integrated circuits may contain as few as 10 or more than 100 000 active devices (transistors and diodes). With the excep-

Table 5.13   Characteristics of some common types of junction gate FET

Device	Type	I_D max. (mA)	V_{DS} max. (V)	P_D max. (mW)	g_{fs} min. (mS)	Application
2N3819	N-chan.	10	25	200	4	General purpose
2N5457	N-chan.	10	25	310	1	General purpose
BF244A	N-chan.	100	30	360	3	RF amplifier

Figure 5.31   N-channel FET test circuit (the arrangement for a P-channel transistor is similar but all meters and supplies must be reversed)

Figure 5.32 Some common transistor packages

operational amplifiers (see Chapter 8) whereas typical examples of digital integrated are logic gates (see Chapter 9). A number of devices bridge the gap between the analogue and digital world. Such devices include analogue to digital converters (ADC), digital to analogue converters (DAC), and timers. For example, the ubiquitous 555 timer contains two operational amplifier stages (configured as voltage comparators) together with a digital bistable stage, a buffer amplifier and an open-collector transistor.

IC packages

As with transistors, a variety of different packages are used for integrated circuits. The most popular form of encapsulation used for integrated circuits is the dual-in-line (DIL) package which may be fabricated from either plastic or ceramic material (with the latter using a glass hermetic sealant) Common DIL packages have 8, 14, 16, 28 and 42 pins on a 0.1 inch matrix.

Flat package (flatpack) construction (featuring both glass–metal and glass–ceramic seals and welded construction) are popular for planar mounting on flat circuit boards. No holes are required to accommodate the leads of such devices which are arranged on a 0.05 inch pitch (i.e. half the pitch used with DIL devices). Single-in-line (SIL) and quad-in-line (QIL) packages are also becoming increasingly popular while TO5, TO72, TO3 and TO220 encapsulations are also found (the latter being commonly used for three-terminal voltage regulators), see Fig. 5.33.

tion of a few specialized applications (such as amplification at high-power levels) integrated circuits have largely rendered a great deal of conventional discrete circuitry obsolete.

Integrated circuits can be divided into two general classes, **linear** (analogue) and **digital**. Typical examples of linear integrated circuits are

Figure 5.33 Some common integrated circuit packages

Circuit symbols introduced in this chapter

Figure 5.34

Formulae introduced in this chapter

LED series resistor:
(page 88)

$R = (V - V_F)/I$

Bipolar transistor currents:
(page 93)

$I_E = I_B + I_C$

Static current gain for a bipolar transistor:
(page 93)

$h_{FE} = I_C/I_B$

Small signal current gain for a bipolar transistor:
(page 94)

$h_{fe} = I_c/I_b$

Forward transfer conductance for a FET:
(page 99)

$g_{fs} = I_d/V_{gs}$

Problems

5.1   Figure 5.35 shows the characteristics of a diode. What type of material is used in this diode? Give a reason for your answer.

5.2   Use the characteristic shown in Fig. 5.35 to determine the resistance of the diode when (a) $V_F = 0.65\,V$ and (b) $I_F = 4\,mA$.

5.3   The following data refers to a signal diode:

V_F (V):   0.0 0.1   0.2   0.3   0.4   0.5   0.6   0.7
I_F (mA): 0.0 0.05 0.02 1.2   3.6   6.5 10.1 13.8

Plot the characteristic and use it to determine:

Forward current (mA)

Forward voltage (V)

Figure 5.35

(a) the forward current when $V_F = 350\,\text{mV}$;
(b) the forward voltage when $I_F = 15\,\text{mA}$.

5.4   A diode is marked 'BZY88C9V1'. What type of diode is it? What is its rated voltage? State one application for the diode.

5.5   An LED is to be used to indicate the presence of a 5 V d.c. supply. If the LED has a nominal forward voltage of 2 V, and is rated at a current of 12 mA, determine the value of series resistor required.

5.6   Identify each of the following transistors:

(a) AF117   (b) BC184
(c) BD131   (d) BF180.

5.7   A transistor operates with a collector current of 2.5 A and a base current 125 mA. Determine the value of emitter current and static common-emitter current gain.

5.8   A transistor operates with a collector current of 98 mA and an emitter current of 103 mA. Determine the value of base current and the static value of common-emitter current gain.

5.9   A bipolar transistor is to be used in a driver circuit in which a base current of 12 mA is available. If the load requires a current of

Figure 5.36

Figure 5.37

200 mA, determine the minimum value of common-emitter current gain required.

5.10   An NPN transistor is to operate with $V_{CE} = 10\,V$, $I_C = 50\,mA$, and $I_B = 400\,\mu A$. Which of the devices listed in Table 5.11 is most suitable for use in this application.

5.11   A transistor is used in a linear amplifier arrangement. The transistor has small and large signal current gains of 250 and 220, respectively, and bias is arranged so that the static value of collector current is 2 mA. Determine the value of base bias current and the change of output (collector) current that would result from a 5 μA change in input (base) current.

5.12   The transfer characteristic for an NPN transistor is shown in Fig. 5.36. Use this characteristic to determine:

(a) I_C when $I_B = 50\,\mu A$;
(b) h_{FE} when $I_B = 50\,\mu A$;
(c) h_{fe} when $I_C = 75\,mA$.

5.13   The output characteristic of an NPN transistor is shown in Fig. 5.37. Use this characteristic to determine:

(a) I_C when $I_B = 100\,\mu A$ and $V_{CE} = 4\,V$;
(b) V_{CE} when $I_B = 40\,\mu A$ and $I_C = 5\,mA$;
(c) I_B when $I_C = 7\,mA$ and $V_{CE} = 6\,V$.

5.14   An N-channel FET operates with a drain current of 20 mA and a gate–source bias of $-1\,V$. If the device has a g_{fs} of 8 mS, determine the new drain current if the bias voltage increases to $-1.5\,V$.

5.15   The following results were obtained during an experiment on an N-channel FET:

V_{GS} (V)	I_D (mA)
0	11.4
−2	7.6
−4	3.8
−6	0.2
−8	0

Plot the mutual characteristic for the FET and use it to determine g_{fs} when $I_D = 5\,mA$.

(Answers to these problems appear on page 261.)

6

Power supplies

This chapter deals with the unsung hero of most electronic systems, the power supply. Nearly all electronic circuits require a source of well regulated d.c. at voltages of typically between 5 V and 30 V. In some cases, this supply can be derived directly from batteries (e.g. 6 V, 9 V, 12 V) but in many others it is desirable to make use of a standard a.c. mains outlet. This chapter explains how rectifier and smoothing circuits operate and how power supply output voltages can be closely regulated. The chapter concludes with a brief description of some practical power supply circuits.

The block diagram of a d.c. power supply is shown in Fig. 6.1. Since the mains input is at a relatively high voltage, a step-down transformer of appropriate turns ratio is used to convert this to a low voltage. The a.c. output from the transformer secondary is then rectified using conventional silicon rectifier diodes (see Chapter 5) to produce an unsmoothed (sometimes referred to as **pulsating d.c.**) output. This is then smoothed and filtered

before being applied to a circuit which will **regulate** (or **stabilize**) the output voltage so that it remains relatively constant in spite of variations in both load current and incoming mains voltage.

Figure 6.2 shows how some of the electronic components that we have already met can be used in the realization of the block diagram in Fig. 6.1. The iron-cored stepdown transformer feeds a rectifier arrangement (often based on a bridge circuit). The output of the rectifier is then applied to a high-value **reservoir** capacitor. This capacitor stores a considerable amount of charge and is being constantly topped-up by the rectifier arrangement. The capacitor also helps to smooth out the voltage pulses produced by the rectifier. Finally, a stabilizing circuit (often based on a **series transistor regulator** and a zener diode **voltage reference**) provides a constant output voltage.

We shall now examine each stage of this arrangement in turn, building up to some complete power supply circuits at the end of the chapter.

Figure 6.1   Block diagram of a d.c. power supply

Figure 6.2   Block diagram of a d.c. power supply showing principal components used in each stage

Rectifiers

Semiconductor diodes (see Chapter 5) are commonly used to convert alternating current (a.c.) to direct current (d.c.), in which case they are referred to as **rectifiers**. The simplest form of rectifier circuit makes use of a single diode and, since it operates on only either positive or negative half-cycles of the supply, it is known as a **half-wave** rectifier.

Figure 6.3 shows a simple half-wave rectifier circuit. Mains voltage (240 V) is applied to the primary of a stepdown transformer (T1). The secondary of T1 steps down the 240 V r.m.s. to

Figure 6.3   Simple half-wave rectifier circuit

(a)

(b)

Figure 6.4   (a) Half-wave rectifier circuit with D1 conducting (positive-going half-cycles of secondary voltage) (b) half-wave rectifier circuit with D1 not conducting (negative-going half-cycles of secondary voltage)

12 V r.m.s. (the turns ratio of T1 will thus be 240/12 or 20:1). Diode D1 will only allow the current to flow in the direction shown (i.e. from cathode to anode). D1 will be forward biased during each positive half-cycle (relative to common) and will effectively behave like a closed switch. When the circuit current tries to flow in the opposite direction, the voltage bias across the diode will be reversed, causing the diode to act like an open switch (see Figs 6.4(a) and 6.4(b), respectively).

The switching action of D1 results in a pulsating output voltage which is developed across the load resistor (R_L). Since the mains supply is at 50 Hz, the pulses of voltage developed across R_L will also be at 50 Hz even if only half the a.c. cycle is present. During the positive half-cycle, the diode will drop the 0.6 V to 0.7 V forward threshold voltage normally associated with silicon diodes. However, during the negative half-cycle the peak a.c. voltage will be dropped across D1 when it is reverse biased. This is an important consideration when selecting a diode for a particular application.

Assuming that the secondary of T1 provides 12 V r.m.s., the peak voltage output from the transformer's secondary winding will be given by:

$$V_{pk} = 1.414 \times V_{r.m.s.} = 1.414 \times 12\,V = 16.968\,V$$

The peak voltage applied to D1 will thus be approximately 17 V. The negative half-cycles are blocked by D1 and thus only the positive half-cycles appear across R_L. Note, however, that the actual peak voltage across R_L will be the 17 V positive peak being supplied from the secondary on T1, **minus** the 0.7 V forward threshold voltage dropped by D1. In other words, positive half-cycle pulses having a peak amplitude of 16.3 V will appear across R_L.

Example 6.1

A mains transformer having a turns ratio of 11:1 is connected to a 220 V r.m.s. mains supply. If the secondary output is applied to a half-wave rectifier, determine the peak voltage that will appear across a load.

Solution

The r.m.s. secondary voltage will be given by:

$$V_s = V_p/11 = 220/44 = 5\,V$$

The peak voltage developed after rectification will be given by:

$$V_{PK} = 1.414 \times 20\,V = 7.07\,V$$

Figure 6.5 Half-wave rectifier with reservoir capacitor

Assuming that the diode is a silicon device with a forward voltage drop of 0.7 V, the actual peak voltage dropped across the load will be:

$$V_L = 7.07\,\text{V} - 0.7\,\text{V} = 6.37\,\text{V}$$

Reservoir and smoothing circuits

Figure 6.5 shows a considerable improvement to the circuit of Fig. 6.3. The capacitor, $C1$, has been added to ensure that the output voltage remains at, or near, the peak voltage even when the diode is not conducting. When the primary voltage is first applied to T1, the first positive half-cycle output from the secondary will charge $C1$ to the peak value seen across R_L. Hence $C1$ charges to 16.3 V at the peak of the positive half-cycle. Because $C1$ and R_L are in parallel, the voltage across R_L will be the same as that across $C1$.

The time required for $C1$ to charge to the maximum (peak) level is determined by the charging circuit time constant (the series resistance multiplied by the capacitance value). In this circuit, the series resistance comprises the secondary winding resistance together with the forward resistance of the diode and the (minimal) resistance of the wiring and connections. Hence $C1$ charges very rapidly as soon as D1 starts to conduct.

The time required for $C1$ to discharge is, in contrast, very much greater. The discharge time constant is determined by the capacitance value and the load resistance, R_L. In practice, R_L is very much larger than the resistance of the secondary circuit and hence $C1$ takes an appreciable time to discharge. During this time, D1 will be reverse biased and will thus be held in its non-conducting state. As a consequence, the only discharge path for $C1$ is through R_L.

$C1$ is referred to as a **reservoir** capacitor. It stores charge during the positive half-cycles of secondary voltage and releases it during the negative half-cycles. The circuit of Fig. 6.5 is thus able to

maintain a reasonably constant output voltage across R_L. Even so, $C1$ will discharge by a small amount during the negative half-cycle periods from the transformer secondary. Figure 6.6 shows the secondary voltage waveform together with the voltage developed across R_L with and without $C1$ present. This gives rise to a small variation in the d.c. output voltage (known as **ripple**).

Since ripple is undesirable we must take additional precautions to reduce it. One obvious method of reducing the amplitude of the ripple is that of simply increasing the discharge time constant. This can be achieved either by increasing the value of $C1$ or by increasing the resistance value of R_L. In practice, however, the latter is not really an option because R_L is the effective resistance of the circuit being supplied and we don't usually have the ability to change it! Increasing the value of $C1$ is a more practical alternative and very large capacitor values (often in excess of 4700 μF) are typical.

Figure 6.7 shows a further refinement of the simple power supply circuit. This circuit employs two additional components, $R1$ and $C2$, which act as a filter to remove the ripple. The value of $C2$ is chosen so that the component exhibits a negligible reactance at the ripple frequency (50 Hz for a half-wave rectifier or 100 Hz for a full-wave rectifier – see later). In effect, $R1$ and $C2$ act like a potential divider. The amount of ripple is reduced by an approximate factor equal to:

$$\frac{X_C}{\sqrt{(R1^2 + X_C^2)}}$$

Example 6.2

The R–C smoothing filter in a 50 Hz mains operated half-wave rectifier circuit consists of

Figure 6.6 Waveforms for half-wave rectifier circuits

Figure 6.7 Half-wave rectifier with R–C smoothing circuit

$R1 = 100\,\Omega$ and $C2 = 1000\,\mu\mathrm{F}$. If $1\,\mathrm{V}$ of ripple appears at the input of the circuit, determine the amount of ripple appearing at the output.

Solution

First we must determine the reactance of the capacitor, $C2$, at the ripple frequency ($50\,\mathrm{Hz}$):

$$X_C = \frac{1}{2\pi fC} = \frac{1}{6.28 \times 50 \times 1000 \times 10^{-6}}$$

$$= \frac{1000}{6.28 \times 50} = \frac{1000}{314}$$

Thus $X_C = 3.18\,\Omega$.

The amount of ripple at the output of the circuit (i.e. appearing across $C2$) will be given by:

$$V_{\mathrm{ripple}} = 1\,\mathrm{V} \times \frac{X_C}{\sqrt{(R1^2 + X_C^2)}}$$

$$= 1\,\mathrm{V} \times \frac{3.18}{\sqrt{(100^2 + 3.18^2)}}$$

$$= 0.032\,\mathrm{V} = 32\,\mathrm{mV}$$

Improved ripple filters

A further improvement can be achieved by using an inductor ($L1$) instead of a resistor ($R1$) in the smoothing circuit. This circuit also offers the advantage that the minimum d.c. voltage is dropped across the inductor (in the circuit of Fig. 6.7, the d.c. output voltage is *reduced* by an amount equal to the voltage drop across $R1$).

Figure 6.8 shows the circuit of a half-wave power supply with an L–C smoothing circuit. At the ripple frequency, $L1$ exhibits a high value of inductive reactance while $C1$ exhibits a low value of capacitive reactance. The combined effect is that of an attenuator which greatly reduces the amplitude of the ripple while having a negligible effect on the direct voltage.

Example 6.3

The L–C smoothing filter in a $50\,\mathrm{Hz}$ mains operated half-wave rectifier circuit consists of $L1 = 10\,\mathrm{H}$ and $C2 = 1000\,\mu\mathrm{F}$. If $1\,\mathrm{V}$ of ripple appears at the input of the circuit, determine the amount of ripple appearing at the output.

Solution

Once again, the reactance of the capacitor, $C2$, is $3.18\,\Omega$ (see Example 6.2).

The reactance of $L1$ at $50\,\mathrm{Hz}$ can be calculated from:

$$X_L = 2\pi fL = 2 \times 3.14 \times 50 \times 10 = 3140\,\Omega$$

Figure 6.8 Half-wave rectifier with L–C smoothing circuit

The amount of ripple at the output of the circuit (i.e. appearing across $C2$) will be approximately given by:

$$V_{ripple} = 1\,V \times \frac{X_C}{X_L + X_C}$$
$$= 1\,V \times \frac{3.18}{3142 + 3.18}$$
$$\approx 0.001\,V \text{ or } 1\,mV$$

It is worth comparing this value with that obtained from the previous example!

Full-wave rectifiers

The half-wave rectifier circuit is relatively inefficient as conduction takes place only on alternate half-cycles. A better rectifier arrangement would make use of both positive **and** negative half-cycles. These full-wave rectifier circuits offer a considerable improvement over their half-wave counterparts. They are not only more efficient but are significantly less demanding in terms of the reservoir and smoothing components. There are two basic forms of full-wave rectifier; the bi-phase type and the bridge rectifier type.

Bi-phase rectifier circuits

Figure 6.9 shows a simple bi-phase rectifier circuit. Mains voltage (240 V) is applied to the primary of a stepdown transformer (T1) which has two identical secondary windings, each providing 12 V r.m.s.

Figure 6.9 Bi-phase rectifier

(the turns ratio of T1 will thus be 240/12 or 20:1 for *each* secondary winding).

On positive half-cycles, point A will be positive with respect to point B. Similarly, point B will be positive with respect to point C. In this condition D1 will allow conduction (its anode will be positive with respect to its cathode) while D2 will not allow conduction (its anode will be negative with respect to its cathode). Thus D1 alone conducts on positive half-cycles.

On negative half-cycles, point C will be positive with respect to point B. Similarly, point B will be positive with respect to point A. In this condition D2 will allow conduction (its anode will be positive with respect to its cathode) while D1 will not allow conduction (its anode will be negative with respect to its cathode). Thus D2 alone conducts on negative half-cycles.

Figure 6.10 shows the bi-phase rectifier circuit with the diodes replaced by switches. In Fig. 6.10(a) D1 is shown conducting on a positive half-cycle while in Fig. 6.10(b) D2 is shown conducting. The result is that current is routed through the load *in the same direction* on successive half-cycles. Furthermore, this current is derived alternately from the two secondary windings.

As with the half-wave rectifier, the switching action of the two diodes results in a pulsating output voltage being developed across the load resistor (R_L). However, unlike the half-wave circuit the pulses of voltage developed across R_L will occur at a frequency of 100 Hz (**not** 50 Hz). This doubling of the ripple frequency allows us to use smaller values of reservoir and smoothing capacitor to obtain the same degree of ripple reduction (recall that the reactance of a capacitor is reduced as frequency increases).

As before, the peak voltage produced by each of the secondary windings will be approximately 17 V and the peak voltage across R_L will be 16.3 V (i.e. 17 V less the 0.7 V forward threshold voltage dropped by the diodes).

Figure 6.11 shows how a reservoir capacitor ($C1$) can be added to ensure that the output voltage remains at, or near, the peak voltage even when the diodes are not conducting. This component operates in exactly the same way as for the half-wave circuit, i.e. it charges to approximately 16.3 V at the peak of the positive half-cycle and holds the voltage at this level when the diodes are in their non-conducting states.

The time required for $C1$ to charge to the maximum (peak) level is determined by the charging circuit time constant (the series resistance

(a)

(b)

Figure 6.10 (a) Bi-phase rectifier with D1 conducting and D2 non-conducting. (b) Bi-phase rectifier with D2 conducting and D1 non-conducting

Figure 6.11 Bi-phase rectifier with reservoir capacitor

multiplied by the capacitance value). In this circuit, the series resistance comprises the secondary winding resistance together with the forward resistance of the diode and the (minimal) resistance of the wiring and connections. Hence $C1$ charges very rapidly as soon as either D1 or D2 starts to conduct.

The time required for $C1$ to discharge is, in contrast, very much greater. The discharge time contrast is determined by the capacitance value and the load resistance, R_L. In practice, R_L is very much larger than the resistance of the secondary circuit and hence $C1$ takes an appreciable time to discharge. During this time, D1 and D2 will be reverse biased and held in a non-conducting state. As a consequence, the only discharge path for $C1$ is through R_L. Figure 6.12 shows voltage waveforms for the bi-phase rectifier, with and without $C1$ present.

Bridge rectifier circuits

An alternative to the use of the bi-phase circuit is that of using a four-diode bridge rectifier (see Fig. 6.13) in which opposite pairs of diode conduct on alternate half-cycles. This arrangement avoids the need to have two separate secondary windings.

A full-wave bridge rectifier arrangement is shown in Fig. 6.14. Mains voltage (240 V) is applied to the primary of a stepdown transformer (T1). The secondary winding provides 12 V r.m.s. (approximately 17 V peak) and has a turns ratio of 20:1, as before. On positive half-cycles, point A will be positive with respect to point B. In this condition D1 and D2 will allow conduction while D3 and D4 will not allow conduction. Conversely, on negative half-cycles, point B will be positive with respect to point A. In this condition D3 and D4 will allow conduction while D1 and D2 will not allow conduction.

Figure 6.15 shows the bridge rectifier circuit with the diodes replaced by switches. In Fig. 6.15(a) D1 and D2 are conducting on a positive half-cycle while in Fig. 6.15(b) D3 and D4 are conducting. Once again, the result is that current is routed through the load *in the same direction* on successive half-cycles.

As with the bi-phase rectifier, the switching action of the two diodes results in a pulsating output voltage being developed across the load resistor (R_L). Once again, the peak output voltage is approximately 16.3 V (i.e. 17 V less the 0.7 V forward threshold voltage).

Figure 6.16 shows how a reservoir capacitor ($C1$) can be added to ensure that the output

T1 secondary voltage +17V / 0V / -17V

Voltage across R_L without $C1$ present +16.3V / 0V

Voltage across R_L with $C1$ present Ripple / 0V

Figure 6.12   Waveforms for the bi-phase rectifier

Figure 6.13   Four diodes connected as a bridge rectifier

Figure 6.14   Full-wave bridge rectifier circuit

Figure 6.15   (a) Bridge rectifier with D1 and D2 conducting, D3 and D4 non-conducting (b) bridge rectifier with D1 and D2 non-conducting, D3 and D4 conducting

voltage remains at, or near, the peak voltage even when the diodes are not conducting. This component operates in exactly the same way as for the bi-phase circuit, i.e. it charges to approximately 16.3 V at the peak of the positive half-cycle and holds the voltage at this level when the diodes are in their non-conducting states. This component operates in exactly the same way as for the bi-phase circuit and the voltage waveforms are identical to those shown in Fig. 6.12.

Finally, R–C and L–C ripple filters can be added to bi-phase and bridge rectifier circuits in exactly the same way as those shown for the half-wave rectifier arrangement (see Figs 6.7 and 6.8).

Figure 6.16   Bridge rectifier with reservoir capacitor

Voltage regulators

A simple voltage regulator is shown in Fig. 6.17. R_s is included to limit the zener current to a safe value when the load is disconnected. When a load (R_L) is connected, the zener current (I_Z) will fall as current is diverted into the load resistance (it is usual to allow a minimum current of 2 mA to 5 mA in order to ensure that the diode regulates). The output voltage (V_O) will remain at the zener voltage until regulation fails at the point at which the potential divider formed by R_S and R_L produces a lower output voltage that is less than V_Z. The ratio of R_S to R_L is thus important.

At the point at which the circuit just begins to fail to regulate:

$$V_Z = V_{IN} \times \frac{R_L}{R_L + R_S}$$

where V_{IN} is the unregulated input voltage. Thus the *maximum* value for R_S can be calculated from:

$$R_S \max = R_L \times \left(\frac{V_{IN}}{V_Z} - 1 \right)$$

The power dissipated in the zener diode, $P_Z = I_Z \times V_Z$, hence the minimum value for R_S can be determined from the 'off-load' condition when:

$$R_S \min = \frac{V_{IN} - V_Z}{I_Z} = \frac{V_{IN} - V_Z}{(P_Z \max / V_Z)}$$
$$= \frac{(V_{IN} - V_Z) \times V_Z}{P_Z \max}$$

thus

$$R_S \min = \frac{(V_{IN} \times V_Z) - V_Z^2}{P_Z \max}$$

where $P_Z \max$ is the maximum rated power dissipation for the zener diode.

Example 6.4

A 5 V zener diode has a maximum rated power dissipation of 500 mW. If the diode is to be used in a simple regulator circuit to supply a regulated 5 V to a load having a resistance of 400 Ω, determine a suitable value of series resistor for operation in conjunction with a supply of 9 V.

Solution

First we should determine the maximum value for R_s:

$$R_s \max = R_L \times \left(\frac{V_{IN}}{V_Z} - 1 \right)$$

thus

$$R_s \max = 400\,\Omega \times \left(\frac{9\,V}{5\,V} - 1 \right) = 400 \times (1.8 - 1)$$
$$= 320\,\Omega$$

Now determine the minimum value for R_s:

$$R_s \min = \frac{(V_{IN} \times V_Z) - V_Z^2}{P_Z \max} = \frac{(9 \times 5) - 5^2}{0.5}$$
$$= \frac{45 - 25}{0.5} = 40\,\Omega$$

Hence a suitable value for R_s would be 150 Ω (roughly mid-way between the two extremes).

Output resistance and voltage regulation

In a perfect power supply, the output voltage would remain constant regardless of the current taken by the load. In practice, however, the output voltage falls as the load current increases. To account for this fact, we say that the power supply has **internal resistance** (ideally this should be zero).

Figure 6.17   Zener diode shunt voltage regulator

This internal resistance appears at the output of the supply and is defined as the change in output voltage divided by the corresponding change in output current. Hence:

$$R_O = \frac{\text{change in output voltage}}{\text{change in output current}} = \frac{dV_O}{dI_L}$$

(where dV_O represents a small change in output voltage and dI_L represents a corresponding small change in output current)

The **regulation** of a power supply is given by the relationship:

$$\text{regulation} = \frac{\text{change in output voltage}}{\text{change in line (input) voltage}} \times 100\%$$

Ideally, the value of regulation should be very small. Simple shunt zener diode regulators are capable of producing values of regulation of 5% to 10%. More sophisticated circuits based on discrete components produce values of between 1% and 5% and integrated circuit regulators often provide values of 1% or less.

Example 6.5

The following data were obtained during a test carried out on a d.c. power supply:

(i) Load test

Output voltage (no-load) = 12 V
Output voltage (2 A load current) = 11.5 V

(ii) Regulation test

Output voltage (mains input, 220 V) = 12 V
Output voltage (mains input, 200 V) = 11.9 V

Determine (a) the equivalent output resistance of the power supply and (b) the regulation of the power supply.

Solution

The output resistance can be determined from the load test data:

$$R_O = \frac{\text{change in output voltage}}{\text{change in output current}} = \frac{(12 - 11.5\,\text{V})}{(2 - 0)}$$
$$= \frac{0.5}{2} = 0.25\,\Omega$$

The regulation can be determined from the regulation test data:

$$\text{regulation} = \frac{\text{change in output voltage}}{\text{change in line (input) voltage}} \times 100\%$$

thus

$$\text{regulation} = \frac{(12 - 11.9)}{220 - 200)} \times 100\%$$
$$= \frac{0.1}{20} \times 100\% = 0.5\%$$

Practical power supply circuits

Figure 6.18 shows a simple power supply circuit capable of delivering an output current of up to 250 mA. The circuit uses a full-wave bridge rectifier arrangement (D1 to D4) and a simple $C-R$ filter. The output voltage is regulated by the shunt-connected 12 V zener diode.

Figure 6.19 shows an improved power supply in which a transistor is used to provide current gain and minimize the power dissipated in the zener diode (TR1 is sometimes referred to as a **series-pass** transistor). The zener diode, D5, is rated at 13 V and the output voltage will be approximately 0.7 V less than this (i.e. 13 V minus the base–emitter voltage drop associated with TR1). Hence

Figure 6.18   Simple d.c. power supply with shunt zener diode regulated output

Figure 6.19   Improved regulated d.c. power supply with series-pass transistor

Figure 6.20   Variable d.c. power supply

Figure 6.21   Power supply with three-terminal IC voltage regulator

the output voltage is about 12.3 V. The circuit is capable of delivering an output current of up to 500 mA (note that TR1 should be fitted with a small heatsink to conduct away any heat produced).

Figure 6.20 shows a variable power supply. The base voltage to the series-pass transistor is derived from a potentiometer connected across the zener diode, D5. Hence the base voltage is variable from 0 V to 13 V. The transistor requires a substantial heatsink (note that TR1's dissipation *increases* as the output voltage is reduced).

Finally, Fig. 6.21 shows a d.c. power supply based on a fixed-voltage three-terminal integrated circuit voltage regulator. These devices are available in standard voltage and current ratings (e.g. 5 V, 12 V, 15 V at 1 A, 2 A and 5 A) and they provide excellent performance in terms of output resistance, ripple rejection and voltage regulation. In addition, such devices usually incorporate over-current protection and can withstand a direct short-circuit placed across their output terminals. This is an essential feature in many practical applications!

Circuit symbols introduced in this chapter

Figure 6.22

Problems

6.1 A half-wave rectifier is fitted with an $R-C$ smoothing filter comprising $R = 200\,\Omega$ and $C = 50\,\mu F$. If 2 V of 400 Hz ripple appear at the input of the circuit, determine the amount of ripple appearing at the output.

6.2 The $L-C$ smoothing filter fitted to a 50 Hz mains operated full-wave rectifier circuit consists of $L = 4\,H$ and $C = 500\,\mu F$. If 4 V of ripple appear at the input of the circuit, determine the amount of ripple appearing at the output.

6.3 If a 9 V zener diode is to be used in a simple shunt regulator circuit to supply a load having a nominal resistance of $300\,\Omega$, determine the maximum value of series resistor for operation in conjunction with a supply of 15 V.

6.4 The circuit of a d.c. power supply is shown in Fig. 6.23. Determine the voltages that will appear at test points A, B and C.

6.5 In Fig. 6.23, determine the current flowing in $R1$ and the power dissipated in D5 when the circuit is operated without any load connected.

6.6 In Fig. 6.23, determine the effect of each of the following fault conditions:

(a) $R1$ open-circuit;
(b) D5 open-circuit;
(c) D5 short-circuit.

6.7 The following data were obtained during a load test carried out on a d.c. power supply:

Output voltage (no-load) = 8.5 V
Output voltage (800 mA load current) = 8.1 V

Determine the output resistance of the power supply and estimate the output voltage at a load current of 400 mA.

6.8 The following data were obtained during a regulation test on a d.c. power supply:

Output voltage (mains input, 230 V) = 15 V
Output voltage (mains input, 190 V) = 14.6 V

Determine the regulation of the power supply and estimate the output voltage when the input voltage is 245 V.

(Answers to these problems appear on page 261.)

Figure 6.23

7

Amplifiers

This chapter introduces the basic concepts of amplifiers and amplification. It describes the most common types of amplifier and outlines the basic classes of operation used in both linear and non-linear amplifiers. The chapter also describes methods for predicting the performance of an amplifier based on equivalent circuits and on the use of semiconductor characteristics and load-lines. Once again, we conclude with a selection of practical circuits that can be built and tested.

Types of amplifier

Many different types of amplifier are found in electronic circuits. Before we explain the operation of transistor amplifiers in detail, it is worth describing some of the types of amplifier used in electronic circuits:

a.c. coupled amplifiers

In a.c. coupled amplifiers, stages are coupled together in such a way that d.c. levels are isolated and only the a.c. components of a signal are transferred from stage to stage.

d.c. coupled amplifiers

In d.c. (or direct) coupled, stages are coupled together in such a way that stages are not isolated to d.c. poentials. Both a.c. and d.c. signal components are transferred from stage to stage.

Large-signal amplifiers

Large-signal amplifiers are designed to cater for appreciable voltage and/or current levels (typically from 1 V to 100 V or more).

Small-signal amplifiers

Small-signal amplifiers are designed to cater for low level signals (normally less than 1 V and often much smaller).

Audio frequency amplifiers

Audio frequency amplifiers operate in the band of frequencies that is normally associated with audio signals (e.g. 20 Hz to 20 kHz).

Wideband amplifiers

Wideband amplifiers are capable of amplifying a very wide range of frequencies, typically from a few tens of hertz to several megahertz.

Radio frequency amplifiers

Radio frequency amplifiers operate in the band of frequencies that is normally associated with radio signals (e.g. from 100 kHz to over 1 GHz). Note that it is desirable for amplifiers of this type to be **frequency selective** and thus their frequency response may be restricted to a relatively narrow band of frequencies (see Fig. 7.10).

Low-noise amplifiers

Low-noise amplifiers are designed so that they contribute negligible noise (signal disturbance) to the signal being amplified. These amplifiers are usually designed for use with very small signal levels (usually less than 10 mV or so).

Gain

One of the most important parameters of an amplifier is the amount of amplification or **gain** that it provides. Gain is simply the ratio of output voltage to input voltage, output current to input current, or output power to input power (see Fig. 7.1). These three ratios give, respectively, the voltage gain, current gain and power gain. Thus:

voltage gain, $A_V = \dfrac{V_{out}}{V_{in}}$

current gain, $A_i = \dfrac{I_{out}}{I_{in}}$

Figure 7.1   Block diagram for an amplifier showing input and output voltages and currents

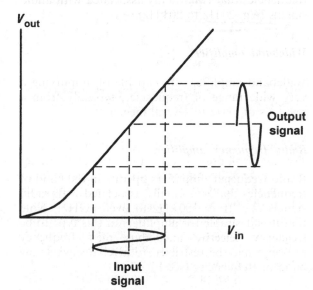

Figure 7.2   Class A (linear) operation

and

power gain, $A_p = \dfrac{P_{out}}{P_{in}}$

Note that, since power is the product of current and voltage ($P = IV$), we can infer that:

$$A_p = \frac{P_{out}}{P_{in}} = \frac{I_{out} \times V_{out}}{I_{in} \times V_{in}} = \frac{I_{out}}{I_{in}} \times \frac{V_{out}}{V_{in}} = A_i \times A_v$$

Hence,

$A_p = A_i \times A_v$

Example 7.1

An amplifier produces an output voltage of 2 V for an input of 50 mV. If the input and output currents in this condition are, respectively, 4 mA and 200 mA, determine:

(a) the voltage gain;
(b) the current gain;
(c) the power gain.

Solution

(a) The voltage gain is calculated from:

$$A_v = \frac{V_{out}}{V_{in}} = \frac{2\,\text{V}}{50\,\text{mV}} = 40$$

(b) The current gain is calculated from:

$$A_i = \frac{I_{out}}{I_{in}} = \frac{200\,\text{mA}}{4\,\text{mA}} = 50$$

(c) The power gain is calculated from:

$$A_p = \frac{P_{out}}{P_{in}} = \frac{V_{out} \times I_{out}}{V_{in} \times I_{in}}$$

$$= \frac{2\,\text{V} \times 200\,\text{mA}}{50\,\text{mV} \times 4\,\text{mA}} = \frac{0.4\,\text{W}}{200\,\mu\text{W}}$$

thus

$A_p = 2000$

(Note that $A_p = A_v \times A_i = 50 \times 40 = 2000$.)

Class of operation

A requirement of most amplifiers is that the output signal should be a faithful copy of the input signal, albeit somewhat larger in amplitude. Other types of amplifier are 'non-linear', in which case their input and output waveforms will not necessarily be similar. In practice, the degree of linearity provided by an amplifier can be affected by a number of factors including the amount of bias applied (see page 118) and the amplitude of the input signal. It is also worth noting that a linear amplifier will become non-linear when the applied input signal exceeds a threshold value. Beyond this value the amplifier is said to be 'over-driven' and the output will become increasingly distorted if the input signal is further increased.

Amplifiers are usually designed to be operated with a particular value of bias supplied to the active devices (i.e. transistors). For linear operation, the active device(s) must be operated in the linear part of their transfer characteristic (V_{out} plotted against V_{in}). In Fig. 7.2 the input and output signals for an amplifier are operating in linear mode. This form of operation is known as

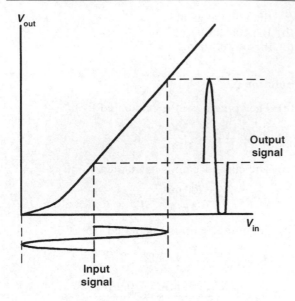

Figure 7.3 Effect of reducing bias and increasing input signal amplitude (the output waveform is no longer a faithful reproduction of the input)

Figure 7.4 Class AB operation (bias set at projected cut-off)

Class A and the bias point is adjusted to the mid-point of the linear part of the transfer characteristic. Furthermore, current will flow in the active devices used in a Class A amplifier during a complete cycle of the signal waveform. At no time does the current fall to zero.

Figure 7.3 shows the effect of moving the bias point down the transfer characteristic and, at the same time, increasing the amplitude of the input signal. From this, you should notice that the extreme negative portion of the output signal has become distorted. This effect arises from the non-linearity of the transfer characteristic that occurs near the origin (i.e. the zero point). Despite the obvious non-linearity in the output waveform, the active device(s) will conduct current during a complete cycle of the signal waveform.

Now consider the case of reducing the bias even further while further increasing the amplitude of the input signal (see Fig. 7.4). Here the bias point has been set at the **projected cut-off** point. The negative-going portion of the output signal becomes cut off (or **clipped**) and the active device(s) will cease to conduct for this part of the cycle. This mode of operation is known as **Class AB**.

Now let's consider what will happen if no bias at all is applied to the amplifier (see Fig. 7.5). The output signal will only comprise a series of positive-going half-cycles and the active device(s) will only

be conducting during half-cycles of the waveform (i.e. they will only be operating 50% of the time). This mode of operation is known as **Class B** and is commonly used in **push–pull** power amplifiers where the two active devices in the output stage operate on alternate half-cycles of the waveform.

Finally, there is one more class of operation to consider. The input and output waveforms for **Class C** operation are shown in Fig. 7.6. Here the bias point is set at beyond the cut-off (zero) point and a very large input signal is applied. The output waveform will then comprise a series of quite sharp positive-going pulses. These pulses of current or voltage can be applied to a tuned circuit load in order to recreate a sinusoidal signal. In effect, the pulses will excite the tuned circuit and

Figure 7.5 Class B operation (no bias applied)

Figure 7.6 Class C operation (bias is set beyond cut-off)

Figure 7.7 Input and output resistances 'seen' looking into the input and output terminals, respectively

its inherent **flywheel action** will produce a sinusoidal output waveform. This mode of operation is only used in RF power amplifiers which must operate at high levels of efficiency.

Table 7.1 summarizes the classes of operation used in amplifiers.

Input and output resistance

Input resistance is the ratio of input voltage to input current and it is expressed in ohms. The input of an amplifier is normally purely resistive (i.e. any reactive component is negligible) in the middle of its working frequency range (i.e. the **mid-band**). In some cases, the reactance of the input may become appreciable (e.g. if a large value of stray capacitance appears in parallel with the input resistance). In such cases we would refer to **input impedance** rather than input resistance.

Output resistance is the ratio of open-circuit output voltage to short-circuit output current and is measured in ohms. Note that this resistance is internal to the amplifier and should not be confused with the resistance of a load connected externally. As with input resistance, the output of an amplifier is normally purely resistive and we

can safely ignore any reactive component. If this is not the case, we would once again refer to **output impedance** rather than output resistance.

Figure 7.7 shows how the input and output resistances are 'seen' looking into the input and output terminals, respectively. We shall be returning to this **equivalent circuit** later in this chapter.

Frequency response

The frequency response of an amplifier is usually specified in terms of the upper and lower **cut-off frequencies** of the amplifier. These frequencies are those at which the output power has dropped to 50% (otherwise known as the **−3 dB points**) or where the voltage gain has dropped to 70.7% of its mid-band value. Figures 7.8 and 7.9, respectively, show how the bandwidth can be expressed in terms of power or voltage. In either case, the cut-off frequencies (f_1 and f_2) and bandwidth are identical.

The frequency response characteristics for various types of amplifier are shown in Fig. 7.10. Note that, for response curves of this type, frequency is almost invariably plotted on a logarithmic scale.

Example 7.2

Determine the mid-band voltage gain and upper and lower cut-off frequencies for the amplifier whose frequency response is shown in Fig. 7.11.

Table 7.1 Classes of operation

Class of operation	Bias point	Conduction angle (typical)	Efficiency (typical)	Application
A	Mid-point	360°	5% to 20%	Linear audio amplifiers
AB	Projected cut-off	210°	20% to 40%	Push–pull audio amplifiers
B	At cut-off	180°	40% to 70%	Push–pull audio amplifiers
C	Beyond cut-off	120°	70% to 90%	Radio frequency power amplifiers

Figure 7.8   Frequency response and bandwidth (output power plotted against frequency)

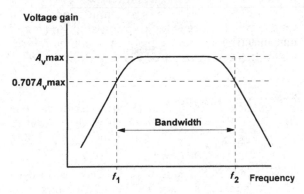

Figure 7.9   Frequency response and bandwidth (output voltage plotted against frequency)

Solution

The mid-band voltage gain corresponds with the flat part of the frequency response characteristic. At the point the voltage gain reaches a maximum of 35 (see Fig. 7.11).

The voltage gain at the cut-off frequencies can be calculated from:

$$A_V \text{ cut-off} = 0.707 \times A_V \text{ max}$$
$$= 0.707 \times 35 = 24.7$$

This value of gain intercepts the frequency response at $f_1 = 57$ Hz and $f_2 = 590$ kHz (see Fig. 7.11).

Bandwidth

The bandwidth of an amplifier is usually taken as the difference between the upper and lower cut-off frequencies (i.e. $f_2 - f_1$ in Figs 7.9 and 7.10). The bandwidth of an amplifier must be sufficient to accommodate the range of frequencies present within the signals that it is to be presented with. Many signals contain **harmonic** components (i.e. signals at $2f$, $3f$, $4f$, etc. where f is the frequency of the **fundamental** signal). To perfectly reproduce a square wave, for example, requires an amplifier with an infinite bandwidth (note that a square wave comprises an infinite series of harmonics). Clearly it is not possible to perfectly reproduce such a wave but it does explain why it can be desirable for an amplifier's bandwidth to greatly exceed the highest signal frequency that it is required to handle!

Phase shift

Phase shift is the phase angle between the input and output voltages measured in degrees. The measurement is usually carried out in the mid-band where, for most amplifiers, the phase shift remains relatively constant. Note also that con-

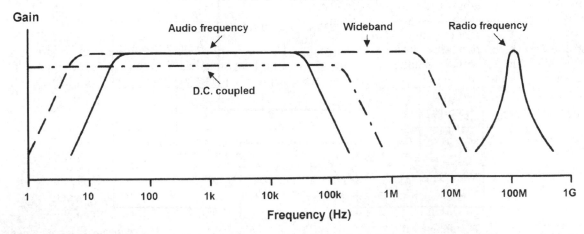

Figure 7.10   Typical frequency response graphs for various types of amplifier

Figure 7.11

ventional single-stage transistor amplifiers usually provide phase shifts of either 180° or 360° (i.e. 0°).

Negative feedback

Many practical amplifiers use negative feedback in order to precisely control the gain, reduce distortion and improve bandwidth. The gain can be reduced to a manageable value by feeding back a small proportion of the output. The amount of feedback determines the overall (or **closed-loop**) gain. Because this form of feedback has the effect of reducing the overall gain of the circuit, this form

of feedback is known as **negative feedback**. An alternative form of feedback, where the output is fed back in such a way as to reinforce the input (rather than to subtract from it) is known as **positive feedback**. This form of feedback is used in oscillator circuits (see Chapter 9).

Figure 7.12 shows the block diagram of an amplifier stage with negative feedback applied. In this circuit, the proportion of the output voltage fed back to the input is given by β and the overall voltage gain will be given by:

$$\text{overall gain} = \frac{V_{\text{out}}}{V_{\text{in}}}$$

Figure 7.12 Amplifier with negative feedback applied

Now $V'_{in} = V_{in} - \beta V_{out}$ (by applying Kirchhoff's voltage law) (note that the amplifier's input voltage has been **reduced** by applying negative feedback) thus

$$V_{in} = V'_{in} + \beta V_{out}$$

and

$$V_{out} = A_V \times V'_{in} \quad (A_V \text{ is the \textbf{internal gain} of the amplifier})$$

Hence,

$$\text{overall gain} = \frac{A_V \times V'_{in}}{V'_{in} + \beta V_{out}} = \frac{A_V \times V'_{in}}{V'_{in} + \beta(A_V \times V'_{in})}$$

Thus

$$\text{overall gain} = \frac{A_V}{1 + \beta A_V}$$

Hence, the overall gain with negative feedback applied will be less than the gain without feedback. Furthermore, if A_V is very large (as is the case with an operational amplifier – see Chapter 8) the overall gain with negative feedback applied will be given by:

overall gain (when A_V is very large) $= 1/\beta$

Note, also, that the **loop gain** of a feedback amplifier is defined as the product of β and A_V.

Example 7.3

An amplifier with negative feedback applied has an open-loop voltage gain of 50 and one-tenth of its output is fed back to the input (i.e. $\beta = 0.1$). Determine the overall voltage gain with negative feedback applied.

Solution

With negative feedback applied the overall voltage gain will be given by:

$$\frac{A_V}{1 + \beta A_V} = \frac{50}{1 + (0.1 \times 50)} = \frac{50}{1 + 5} = \frac{50}{6} = 8.33$$

Example 7.4

If, in Example 7.3, the amplifier's open-loop voltage gain increases by 20%, determine the percentage increase in overall voltage gain.

Solution

The new value of open-loop gain will be given by:

$$A_V = A_V + 0.2 A_V = 1.2 \times 50 = 60$$

The overall voltage gain with negative feedback will then be:

$$\frac{A_V}{1 + \beta A_V} = \frac{60}{1 + (0.1 \times 60)} = \frac{60}{1 + 6} = \frac{50}{6} = 8.57$$

The increase in overall voltage gain, expressed as a percentage, will thus be:

$$\frac{8.57 - 8.33}{8.33} \times 100\% = 2.88\%$$

(Note that this example illustrates one of the important benefits of negative feedback in stabilizing the overall gain of an amplifier stage.)

Example 7.5

An integrated circuit that produces an open-loop gain of 100 is to be used as the basis of an amplifier stage having a precise voltage gain of 20. Determine the amount of feedback required.

Solution

Rearranging the formula, $A_V/(1 + \beta A_V)$, to make β the subject gives:

$$\beta = \frac{1}{A'_V} - \frac{1}{A_V}$$

where A'_V is the overall voltage gain with feedback applied, and A_V is the open-loop voltage gain.

Transistor amplifiers

Regardless of what type of transistor is employed, three basic circuit configurations are used. These three circuit configurations depend upon which one of the three transistor connections is made common to both the input and the output. In the case of bipolar transistors, the configurations are known as **common emitter**, **common collector** (or **emitter follower**) and **common base**. Where field effect transistors are used, the corresponding configurations are **common source**, **common drain** (or **source follower**) and **common gate**.

The three basic circuit configurations (Figs 7.13 to 7.18) exhibit quite different performance characteristics, as shown in Tables 7.2 and 7.3 (typical values are given in brackets).

Figure 7.13 Common-emitter configuration

Figure 7.14 Common-collector (emitter follower) configuration

Figure 7.15 Common-base configuration

Figure 7.16 Common-source configuration

Figure 7.17 Common-drain (source follower) configuration

Figure 7.18 Common-gate configuration

Equivalent circuits

One method of determining the behaviour of an amplifier stage is to make use of an equivalent circuit. Figure 7.19 shows the basic equivalent circuit of an amplifier. The output circuit is reduced

Table 7.2   Bipolar transistor amplifier circuit configurations

Parameter	Mode of operation		
	Common emitter (Fig. 7.13)	Common collector (Fig. 7.14)	Common base (Fig. 7.15)
Voltage gain	medium/high (40)	unity (1)	high (200)
Current gain	high (200)	high (200)	unity (1)
Power gain	very high (8000)	high (200)	high (200)
Input resistance	medium (2.5 kΩ)	high (100 kΩ)	low (200 Ω)
Output resistance	medium/high (20 kΩ)	low (100 Ω)	high (100 kΩ)
Phase shift	180°	0°	0°
Typical applications	General purpose, AF and RF amplifiers	Impedance matching, input and output stages	RF and VHF amplifiers

Table 7.3   Field effect transistor amplifier circuit configurations

Parameter	Mode of operation		
	Common source (Fig. 7.16)	Common drain (Fig. 7.17)	Common gate (Fig. 7.18)
Voltage gain	medium (40)	unity (1)	high (250)
Current gain	very high (200 000)	very high (200 000)	unity (1)
Power gain	very high (8 000 000)	very high (200 000)	high (250)
Input resistance	very high (1 MΩ)	very high (1 MΩ)	low (500 Ω)
Output resistance	medium/high (50 kΩ)	low (200 Ω)	high (150 kΩ)
Phase shift	180°	0°	0°
Typical applications	General purpose, AF and RF amplifiers	Impedance matching stages	RF and VHF amplifiers

Figure 7.19   Simple equivalent circuit for an amplifier

to its Thévenin equivalent (see Chapter 3) comprising a voltage generator ($A_V \times V_{in}$) and a series resistance (R_{out}). This simple model allows us to forget the complex circuitry that might exist within the amplifier box!

In practice, we use a slightly more complex equivalent circuit model in the analysis of a transistor amplifier. The most frequently used equivalent circuit is that which is based on **hybrid**

parameters (or **h-parameters**). In this form of analysis, a transistor is replaced by four components; h_i, h_r, h_f and h_o (see Table 7.4).

In order to indicate which one of the operating modes is used we add a further subscript letter to each h-parameter; e for common emitter, b for common base and c for common collector (see

Table 7.4   Hybrid parameters

h_i input resistance, $\dfrac{\mathrm{d}v_i}{\mathrm{d}i_i}$

h_r reverse voltage transfer ratio, $\dfrac{\mathrm{d}v_i}{\mathrm{d}v_o}$

h_f forward current transfer ratio, $\dfrac{\mathrm{d}i_o}{\mathrm{d}i_i}$

h_o output conductance, $\dfrac{\mathrm{d}i_o}{\mathrm{d}v_o}$

Table 7.5 *h*-parameters for a transistor operating in common-emitter mode

h_{ie}	input resistance,	$\dfrac{dv_{be}}{di_b}$
h_{re}	reverse voltage transfer ratio,	$\dfrac{dv_{be}}{dv_{ce}}$
h_{fe}	forward current transfer ratio,	$\dfrac{di_c}{di_b}$
h_{oe}	output conductance,	$\dfrac{di_c}{dv_{ce}}$

Table 7.5). However, to keep things simple we shall only consider common-emitter operation.

Common-emitter input resistance (h_{ie})

The input resistance of a transistor is the resistance that is effectively 'seen' between its input terminals. As such, it is the ratio of the voltage between the input terminals to the current flowing into the input. In the case of a transistor operating in common-emitter mode, the input voltage is the voltage developed between the base and emitter, V_{be}, while the input current is the current supplied to the base, I_b.

Figure 7.20 shows the current and voltage at the input of a common-emitter amplifier stage while Fig. 7.21 shows how the input resistance, R_{in}, appears between the base and emitter. Note that R_{in} is not a discrete component – it is *inside* the transistor.

From the foregoing we can deduce that:

$$R_{in} = \frac{V_{be}}{I_b}$$

(note that this is the same as h_{ie}).

The transistor's input characteristic can be used to predict the input resistance of a transistor amplifier stage. Since the input characteristic is non-linear (recall that very little happens until the base–emitter voltage exceeds 0.6 V), the value of input resistance will be very much dependent on the exact point on the graph at which the transistor is being operated. Furthermore, we might expect quite different values of resistance according to whether we are dealing with larger d.c. values or smaller incremental changes (a.c. values). Since this can be a rather difficult concept, it is worth expanding on it.

Figure 7.22 shows a typical input characteristic in which the transistor is operated with a base current, I_B, of 50 μA. This current produces a base–emitter voltage, V_{BE}, of 0.65 V. The input resistance corresponding to these steady (d.c.) values will be given by:

$$R_{in} = \frac{V_{BE}}{I_B} = \frac{0.65 \, V}{50 \, \mu A} = 13 \, k\Omega$$

Now, suppose that we apply a steady bias current of, say, 70 μA and superimpose on this a signal

Figure 7.20 Voltage and current at the input of a common-emitter amplifier

Figure 7.21 Input resistance of a common-emitter amplifier stage

Figure 7.22 Using the input characteristic to determine the large-signal (static) input resistance of a transistor connected in common-emitter mode

that varies above and below this value, swinging through a total change of 100 μA (i.e. from 20 μA to 120 μA). Figure 7.23 shows that this produces a base–emitter voltage change of 0.05 V.

The input resistance seen by this small-signal input current is given by:

$$R_{in} = \frac{\text{change in } V_{BE}}{\text{change in } I_B} = \frac{dV_{be}}{dI_b} = \frac{0.05\,V}{100\,\mu A} = 500\,\Omega$$

In other words,

$$h_{ie} = 500\,\Omega \text{ (since } h_{ie} = \frac{dV_{be}}{dI_b})$$

It is worth comparing this value with the steady (d.c.) value. The appreciable difference is entirely attributable to the shape of the input characteristic!

Common-emitter current gain (h_{fe})

The current gain produced by a transistor is the ratio of output current to input current. In the case of a transistor operating in common-emitter mode, the input current is the base current, I_B, while the output current is the collector current, I_C.

Figure 7.24 shows the input and output currents and voltages for a common-emitter amplifier stage. The magnitude of the current produced at the output of the transistor is equal to the current gain, A_i, multiplied by the applied base current, I_b. Since the output current is the current flowing in the collector, I_c, we can deduce that:

$$I_c = A_i \times I_b$$

where $A_i = h_{fe}$ (the common-emitter current gain).

Figure 7.23   Using the input characteristic to determine the small-signal input resistance of a transistor connected in common-emitter mode

Figure 7.24   Input and output currents and voltages in a common-emitter amplifier stage

Figure 7.25   Equivalent output current source in a common-emitter amplifier stage

Figure 7.25 shows how this current source appears between the collector and emitter. Once again, the current source is not a discrete component – it appears *inside* the transistor.

The transistor's transfer characteristic can be used to predict the current gain of a transistor amplifier stage. Since the transfer characteristic is linear, the current gain remains reasonably constant over a range of collector current.

Figure 7.26 shows a typical transfer characteristic in which the transistor is operated with a base current, I_B, of 240 μA. This current produces a collector current, I_C, of 12 mA. The current gain corresponding to these steady (d.c.) values will be given by:

$$A_i = \frac{I_C}{I_B} = \frac{2.5\,mA}{50\,\mu A} = 50$$

(note that this is the same as h_{FE}, the large-signal or steady-state value of common-emitter current gain).

Now, suppose that we apply a steady bias current of, say, 240 μA and superimpose on this a signal that varies above and below this value, swinging through a total change of 220 μA (i.e. from 120 μA to 360 μA). Figure 7.27 shows that this produces a collector current swing of 10 mA.

The small-signal a.c. current gain is given by:

$$A_i = \frac{\text{change in } I_C}{\text{change in } I_B} = \frac{dI_c}{dI_b} = \frac{10\,mA}{220\,\mu A} = 45.45$$

Figure 7.26 Using the transfer characteristic to determine the large-signal (static) current gain of a transistor connected in common-emitter mode

Figure 7.27 Using the transfer characteristic to determine the small-signal current gain of a transistor connected in common-emitter mode

Once again, it is worth comparing this value with the steady-state value (h_{FE}). Since the transfer characteristic is reasonably linear, the values are quite close (45.45 compared with 50). If the transfer characteristic was perfectly linear, then h_{fe} would be equal to h_{FE}.

h-parameter equivalent circuit for common-emitter operation

The complete *h*-parameter equivalent circuit for a transistor operating in common-emitter mode is shown in Fig. 7.28. We have already shown how the two most important parameters, h_{ie} and h_{fe}, can be found from the transistor's characteristics curves.

Figure 7.28 *h*-parameter equivalent circuit for a transistor amplifier

The remaining parameters, h_{re} and h_{oe}, can, in many applications, be ignored. A typical set of *h*-parameters for a BFY50 transistor is shown in Table 7.6. Note how small h_{re} and h_{oe} are.

Example 7.6

A BFY50 transistor is used in a common-emitter amplifier stage with $R_L = 10\,k\Omega$ and $I_C = 1\,mA$. Determine the output voltage produced by an input signal of 10 mV. (You may ignore the effect of h_{re} and any bias components that may be present externally.)

Solution

The equivalent circuit (with h_{re} replaced by a short-circuit) is shown in Fig. 7.29. The load effectively appears between the collector and emitter while the input signal appears between base and emitter.

First we need to find the value of input current, I_b, from:

$$I_b = \frac{V_{in}}{h_{ie}} = \frac{10\,mV}{250\,\Omega} = 40\,\mu A$$

Table 7.6 *h*-parameters for a BFY50 transistor

h_{ie} (Ω)	h_{re}	h_{fe}	h_{oe} (μS)
250	0.85×10^{-4}	80	35

(Measured at $I_C = 1\,mA$, $V_{CE} = 5\,V$.)

Figure 7.29

Next we find the value of current generated, I_f, from:

$$I_f = h_{fe} \times I_b = 80 \times 40\,\mu A = 320\,\mu A$$

This value of current is shared between the internal resistance between collector and emitter (i.e. $1/h_{oe}$) and the external load, R_L.

To determine the value of collector current, we can apply the current divider theorem (Chapter 3):

$$I_c = I_f \times \frac{1/h_{oe}}{1/h_{oe} + R_L}$$

$$= 320\,\mu A \times \frac{1/(80 \times 10^{-6})}{1/(80 \times 10^{-6}) + 10\,k\Omega}$$

Thus

$$I_c = 320\,\mu A \times \frac{12.5\,k\Omega}{12.5\,k\Omega + 10\,k\Omega}$$

$$= 320\,\mu A \times 0.555 = 177.6\,\mu A$$

Finally, we can determine the output voltage from:

$$V_{out} = I_c \times R_L = 177.6\,\mu A \times 10\,k\Omega = 1.776\,V$$

Voltage gain

We can use hybrid parameters to determine the voltage gain of a transistor stage. We have already shown how the voltage gain of an amplifier stage is given by:

$$A_v = \frac{V_{out}}{V_{in}}$$

In the case of a common-emitter amplifier stage, $V_{out} = V_{ce}$ and $V_{in} = V_{be}$. If we assume that h_{oe} and h_{re} are negligible, then:

$$V_{out} = V_{ce} = I_c \times R_L = I_f \times R_L = h_{fe} \times I_b \times R_L$$

and

$$V_{in} = V_{be} = I_b \times R_{in} = I_b \times h_{ie}$$

Thus

$$A_v = \frac{V_{out}}{V_{in}} = \frac{h_{fe} \times I_b \times R_L}{I_b \times h_{ie}} = \frac{h_{fe} \times R_L}{h_{ie}}$$

Example 7.7

A transistor has $h_{fe} = 150$ and $h_{ie} = 1.5\,k\Omega$. Assuming that h_{re} and h_{oe} are both negligible,

determine the value of load resistance required to produce a voltage gain of 200.

Solution

Since

$$A_v = \frac{h_{fe} \times R_L}{h_{ie}}, R_L = \frac{A_v \times h_{ie}}{h_{fe}}$$

To produce a gain of 200,

$$R_L = \frac{200 \times 1.5\,k\Omega}{150} = 2\,k\Omega$$

Bias

We stated earlier that the optimum value of bias for a Class A (linear) amplifier is that value which ensures that the active devices are operated at the mid-point of their transfer characteristics. In practice, this means that a static value of collector current will flow even when there is no signal present. Furthermore, the collector current will flow throughout the complete cycle of an input signal (i.e. conduction will take place over an angle of 360°). At no stage will the transistor be **saturated** nor should it be **cut-off**.

In order to ensure that a static value of collector current flows in a transistor, a small current must be applied to the base of the transistor. This current can be derived from the same voltage rail that supplies the collector circuit (via the load). Figure 7.30 shows a simple Class-A common-emitter amplifier circuit in which the base bias resistor, $R1$, and collector load resistor, $R2$, are connected to a common positive supply rail.

Figure 7.30 Basic Class-A common-emitter amplifier stage

The signal is applied to the base terminal of the transistor via a coupling capacitor, $C1$. This capacitor removes the d.c. component of any signal applied to the input terminals and ensures that the base bias current delivered by $R1$ is unaffected by any device connected to the input. $C2$ couples the signal out of the stage and also prevents d.c. current flowing appearing at the output terminals.

In order to stabilize the operating conditions for the stage and compensate for variations in transistor parameters, base bias current for the transistor can be derived from the voltage at the collector (see Fig. 7.31). This voltage is dependent on the collector current which, in turn, depends upon the base current. A negative feedback loop thus exists in which there is a degree of self-regulation. If the collector current increases, the collector voltage will fall and the base current will be reduced. The reduction in base current will produce a corresponding reduction in collector current to offset the original change. Conversely, if the collector current falls, the collector voltage will rise and the base current will increase. This, in turn, will produce a corresponding increase in collector current to offset the original change.

The negative feedback path in Fig. 7.31 provides feedback that involves an a.c. (signal) component as well as the d.c. bias. As a result of the a.c. feedback, there is a slight reduction in signal gain. The signal gain can be increased by removing the a.c. signal component from the feedback path so that only the d.c. bias component is present. This can be achieved with the aid of a bypass capacitor

Figure 7.32 Improved version of Fig. 7.31

as shown in Fig. 7.32. The value of bypass capacitor, $C2$, is chosen so that the component exhibits a very low reactance at the lowest signal frequency when compared with the series base bias resistance, $R2$. The result of this potential divider arrangement is that the a.c. signal component is effectively bypassed to ground.

Figure 7.33 shows an improved form of transistor amplifier in which d.c. negative feedback is used to stabilize the stage and compensate for variations in transistor parameters, component values and temperature changes. $R1$ and $R2$ form a potential divider that determines the d.c. base potential, V_B. The base–emitter voltage (V_{BE}) is the difference between the potentials present at the base (V_B) and emitter (V_E). The potential at the

Figure 7.31 Improvement on the circuit shown in Fig. 7.30 (using negative feedback to bias the transistor)

Figure 7.33 A common-emitter amplifier stage with effective bias stabilization

emitter is governed by the emitter current (I_E). If this current increases, the emitter voltage (V_E) will increase and, as a consequence V_{BE} will fall. This, in turn, produces a reduction in emitter current which largely offsets the original change. Conversely, if the emitter current decreases, the emitter voltage (V_E) will decrease and V_{BE} will increase (remember that V_B remains constant). The increase in bias results in an increase in emitter current compensating for the original change.

Example 7.8

Determine the static value of current gain and collector voltage in the circuit shown in Fig. 7.34.

Solution

Since 2 V appears across $R4$, we can determine the emitter current easily from:

$$I_E = \frac{V_E}{R4} = \frac{2\,V}{1\,k\Omega} = 2\,mA$$

Next we should determine the base current. This is a little more difficult. The base current is derived from the potential divider formed by $R1$ and $R2$. The potential at the junction of $R1$ and $R2$ is 2.6 V hence we can determine the currents through $R1$ and $R2$, the difference between these currents will be equal to the base current.

The current in $R2$ will be given by:

$$I_{R2} = \frac{V_B}{R2} = \frac{2.6\,V}{33\,k\Omega} = 79\,\mu A$$

The current in $R1$ will be given by:

$$I_{R1} = \frac{9\,V - V_B}{R1} = \frac{6.4\,V}{68\,k} = 94.1\,\mu A$$

Hence base current, $I_B = 94.1\,\mu A - 79\,\mu A = 15.1\,\mu A$.

Next we can determine the collector current from:

$$I_C = I_B + I_C = 15.1\,\mu A + 2\,mA = 2.0151\,mA$$

The current gain can then be determined from:

$$h_{FE} = \frac{I_C}{I_B} = \frac{2.0151\,mA}{15.1\,\mu A} = 133.45$$

Finally we can determine the collector voltage by subtracting the voltage dropped across $R3$ from the 9 V supply. The voltage dropped across $R4$ will be:

$$V_{R4} = I_C \times R4 = 2.0151\,mA \times 2.2\,k\Omega = 4.433\,V$$

Hence collector voltage, $V_C = 9\,V - 4.433\,V = 4.567\,V$.

Predicting amplifier performance

The a.c. performance of an amplifier stage can be predicted using a load line superimposed on the relevant set of output characteristics. For a bipolar transistor operating in common-emitter mode the required characteristics are I_C plotted against V_{CE}. One end of the load line corresponds to the supply voltage (V_{CC}) while the other end corresponds to the value of collector or drain current that would flow with the device totally saturated. In this condition:

$$I_C = \frac{V_{CC}}{R_L}$$

where R_L is the value of collector or drain load resistance.

Figure 7.35 shows a load line superimposed on a set of output characteristics for a bipolar transistor operating in common-emitter mode. The quiescent point (or operating point) is the point on the load line that corresponds to the conditions that exist when no-signal is applied to the stage. In Fig. 7.35, the base bias current is set at 20 μA so that the quiescent point effectively sits roughly halfway along the load line. This position ensures that the collector voltage can swing both positively (above) and negatively (below) its quiescent value (V_{CQ}). The effect of superimposing an alternating base current (of 20 μA peak–peak) to the d.c. bias current (of 20 μA) can be clearly seen. The

Figure 7.34

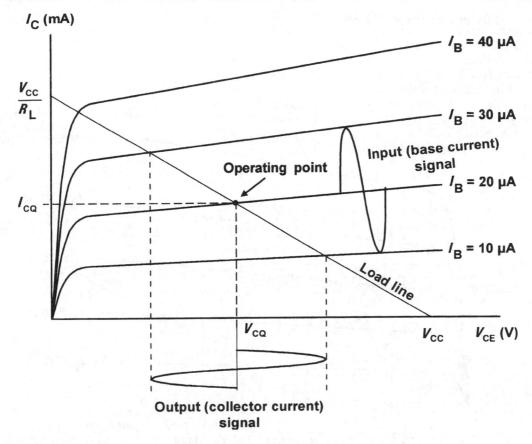

Figure 7.35 Operating point and quiescent values shown on the load line for a bipolar transistor operating in common-emitter mode

corresponding collector current signal can be determined by simply moving up and down the load line.

Example 7.9

The characteristic curves shown in Fig. 7.36 relate to a transistor operating in common-emitter mode. If the transistor is operated with $I_B = 30\,\mu A$, a load resistor of $1.2\,k\Omega$ and an 18 V supply, determine the quiescent values of collector voltage and current (V_{CQ} and I_{CQ}). Also determine the peak–peak output voltage that would be produced by an input signal of $40\,\mu A$ peak–peak.

Solution

First we need to construct the load line. The two ends of the load line will correspond to V_{CC} (18 V) on the collector–emitter voltage axis and

V_{CC}/R_L (18 V/1.2 kΩ or 15 mA) on the collector current axis. Next we locate the operating point (or quiescent point) from the point of intersection of the $I_B = 30\,\mu A$ characteristic and the load line.

Having located the operating point we can read off the quiescent (no-signal) values of collector–emitter voltage (V_{CQ}) and collector current (I_{CQ}). Hence:

$V_{CQ} = 9.2\,V$ and $I_{CQ} = 7.3\,mA$

Next we can determine the maximum and minimum values of collector–emitter voltage by locating the appropriate intercept points on Fig. 7.36. Note that the maximum and minimum values of base current will be $(30\,\mu A + 20\,\mu A)$ on positive peaks of the signal and $(30\,\mu A - 20\,\mu A)$ on negative peaks of the signal.

The maximum and minimum values of V_{CE} are, respectively, 14.8 V and 3.3 V. Hence the output voltage swing will be (14.8 V − 3.3 V) or 11.5 V pk–pk.

Figure 7.36

Practical amplifier circuits

The simple common-emitter amplifier stage shown in Fig. 7.37 provides a modest voltage gain (80 to 120 typical) with an input resistance of approximately 1.5 kΩ and an output resistance of around 20 kΩ. The frequency response extends from a few hertz to several hundred kilohertz.

The improved arrangement shown in Fig. 7.38 provides a voltage gain of around 150 to 200 by eliminating the signal-frequency negative feedback that occurs through R1 in Fig. 7.37.

Figure 7.39 shows a practical common-emitter amplifier with bias stabilization. This stage provides a gain of 150 to well over 200 (depending upon the current gain, h_{fe}, of the individual transistor used). The circuit will operate with supply voltages of between 6 V and 18 V.

Two practical emitter-follower circuits are shown in Figs 7.40 and 7.41. These circuits offer a voltage gain of unity (1) but are ideal for matching a high resistance source to a low resistance load. It is important to note that the input resistance varies with the load connected to the output of the circuit (it is typically in the range 50 kΩ to 150 kΩ). The input resistance can be calculated by multiplying h_{fe} by the effective resistance of R2 in parallel with the load connected to the output terminals.

Figure 7.41 is an improved version of Fig. 7.40 in which the base current is derived from the potential divider formed by R1 and R2. Note, however, that the input resistance is reduced since R1 and R2 effectively appear in parallel with the input. The input resistance of the stage is thus typically in the region of 40 kΩ to 70 kΩ.

Figure 7.37 A practical common-emitter amplifier stage

Figure 7.38 An improved common-emitter amplifier stage

Figure 7.39 A practical common-emitter amplifier stage with bias stabilization

Figure 7.40   A practical emitter-follower stage

Figure 7.41   An improved emitter-follower stage

Symbols introduced in this chapter

Figure 7.42

Important formulae introduced in this chapter

Voltage gain:
(page 116)

$$A_V = \frac{V_{out}}{V_{in}}$$

Current gain:
(page 116)

$$A_i = \frac{I_{out}}{I_{in}}$$

Power gain:
(page 117)

$$A_p = \frac{P_{out}}{P_{in}}$$

$$A_p = A_i \times A_v$$

Gain with negative feedback applied:
(page 122)

$$A_{VNFB} = \frac{A_V}{1 + \beta A_V}$$

Loop gain:
(page 122)

$$A_{\text{VLOOP}} = \beta A_{\text{v}}$$

Input resistance (hybrid equiv.):
(page 124)

$$h_{\text{i}} = \frac{\mathrm{d}v_{\text{i}}}{\mathrm{d}i_{\text{i}}}$$

Reverse voltage transfer ratio (hybrid equiv.):
(page 124)

$$h_{\text{r}} = \frac{\mathrm{d}v_{\text{i}}}{\mathrm{d}v_{\text{o}}}$$

Forward current transfer ratio (hybrid equiv.):
(page 124)

$$h_{\text{f}} = \frac{\mathrm{d}i_{\text{o}}}{\mathrm{d}i_{\text{i}}}$$

Output conductance (hybrid equiv.):
(page 124)

$$h_{\text{o}} = \frac{\mathrm{d}i_{\text{o}}}{\mathrm{d}v_{\text{o}}}$$

Input resistance (common emitter):
(page 125)

$$h_{\text{ie}} = \frac{\mathrm{d}v_{\text{be}}}{\mathrm{d}i_{\text{b}}}$$

Reverse voltage transfer ratio (common emitter):
(page 125)

$$h_{\text{re}} = \frac{\mathrm{d}v_{\text{be}}}{\mathrm{d}v_{\text{ce}}}$$

Forward current transfer ratio (common emitter):
(page 125)

$$h_{\text{fe}} = \frac{\mathrm{d}i_{\text{c}}}{\mathrm{d}i_{\text{b}}}$$

Output conductance (common emitter):
(page 125)

$$h_{\text{oe}} = \frac{\mathrm{d}i_{\text{c}}}{\mathrm{d}v_{\text{ce}}}$$

Voltage gain (common emitter, assuming h_{re} and can be neglected):
(page 128)

$$A_{\text{v}} = \frac{h_{\text{fe}} \times R_{\text{L}}}{h_{\text{ie}}}$$

Problems

7.1 The following measurements were made during a test on an amplifier:

$$V_{\text{in}} = 250\,\text{mV},\ I_{\text{in}} = 2.5\,\text{mA},\ V_{\text{out}} = 10\,\text{V},$$
$$I_{\text{out}} = 400\,\text{mA}.$$

Figure 7.43

Figure 7.44

Determine:

(a) the voltage gain;
(b) the current gain;
(c) the power gain;
(d) the input resistance.

7.2   An amplifier has a power gain of 25 and identical input and output resistances of $600\,\Omega$. Determine the input voltage required to produce an output of 10 V.

7.3   Determine the mid-band voltage gain and upper and lower cut-off frequencies for the amplifier whose frequency response curve is shown in Fig. 7.43.

7.4   An amplifier with negative feedback applied has an open-loop voltage gain of 250, and 5% of its output is fed back to the input. Determine the overall voltage gain with negative feedback applied.

7.5   An amplifier produces an open-loop gain of 180. Determine the amount of feedback required if it is to be operated with a precise voltage gain of 50.

7.6   A transistor has the following h-parameters:

$h_{ie} = 800\,\Omega$
$h_{re} = $ negligible
$h_{fe} = 120$
$h_{oe} = 50\,\mu\text{S}.$

Figure 7.45

Figure 7.46

If the transistor is to be used as the basis of a common-emitter amplifier stage with $R_L = 12\,k\Omega$, determine the output voltage when an input signal of $2\,mV$ is applied.

7.7   Determine the unknown current and voltages in Fig. 7.44.

7.8   The output characteristics of a bipolar transistor are shown in Fig. 7.45. If this transistor is used in an amplifier circuit operating from a 12 V supply with a base bias current of $60\,\mu A$ and a load resistor of $1\,k\Omega$, determine the quiescent values of collector–emitter voltage and collector current. Also determine the peak–peak output

voltage produced when an $80\,\mu A$ peak–peak signal current is applied.

7.9   The output characteristics of a field effect transistor are shown in Fig. 7.46. If this FET is used in an amplifier circuit operating from an 18 V supply with a gate-source bias voltage of $-3\,V$ and a load resistor of $900\,\Omega$, determine the quiescent values of drain–source voltage and drain current. Also determine the peak–peak output voltage when an input voltage of 2 V pk–pk is applied. Also determine the voltage gain of the stage.

(Answers to these problems can be found on page 261.)

8

Operational amplifiers

This chapter introduces a highly versatile family of analogue integrated circuits. These operational amplifier 'gain blocks' offer near-ideal characteristics (i.e. virtually infinite voltage gain and input resistance coupled with low output resistance and wide bandwidth).

External components are added to operational amplifiers in order to define their function within a circuit. By adding two resistors, we can produce an amplifier having a precisely defined gain. Alternatively, with just one resistor and one capacitor we can produce an active integrating circuit. This chapter introduces the basic concepts of operational amplifiers and describes their use in a number of practical circuit applications.

Symbols and connections

The symbol for an operational amplifier is shown in Fig. 8.1. There are a few things to note about this. The operational amplifier has two input connections and one output connection. There is no direct connection to common. Furthermore, to keep circuits simple we don't always show the connections to the supply – it is often clearer to leave them out of the circuit altogether!

In Fig. 8.1, one of the inputs is marked '−' and the other is marked '+'. These polarity markings have nothing to do with the supply connections – they indicate the overall phase shift between each input and the output. The '+' sign indicates zero phase shift while the '−' sign indicates 180° phase shift. Since 180° phase shift produces an inverted (i.e. turned upside down) waveform, the '−' input

is often referred to as the **inverting input**. Similarly, the '+' input is known as the **non-inverting input**.

Most (but not all) operational amplifiers require a **symmetrical supply** (of typically ±5 V to ±15 V). This allows the output voltage to swing both positive (above 0 V) and negative (below 0 V). Figure 8.2 shows how the supply connections would appear if we decided to include them. Note that we usually have two separate supplies; a positive supply and an equal, but opposite, negative supply. The common connection to these two supplies (i.e. the 0 V rail) acts as the **common rail** in our circuit. The input and output voltages are usually measured relative to this rail. Figure 8.3 shows how the supplies are connected.

Terminology

Before we take a look at some of the characteristics of 'ideal' and 'real' operational amplifiers it is important to define some of the terms that we apply to these devices. Some of these terms (such as voltage gain, input resistance and output resistance) were introduced briefly in Chapter 7. We shall now expand on the definitions of these terms in relation to operational amplifiers.

Figure 8.2   Operational amplifier with supply connections

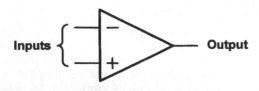

Figure 8.1   Symbol for an operational amplifier

Figure 8.3 A typical operational amplifier power supply arrangement

Open-loop voltage gain

The open-loop voltage gain of an operational amplifier is defined as the ratio of output voltage to input voltage measured with no feedback applied. Open-loop voltage gain may thus be thought of as the 'internal' voltage gain of the device. In practice, this value is exceptionally high (typically greater than 100 000) but is liable to considerable variation from one device to another.

Open-loop voltage gain is the ratio of output voltage to input voltage measured without feedback applied, hence:

$$A_{VOL} = V_{out}/V_{in}$$

where A_{VOL} is the open-loop voltage gain, V_{out} and V_{in} are the output and input voltages, respectively, under open-loop conditions. In linear voltage amplifying applications, a large amount of negative feedback will normally be applied and the open-loop voltage gain can be thought of as the internal voltage gain provided by the device.

The open-loop voltage gain is often expressed in decibels (dB) rather than as a ratio (see Appendix 5). In this case:

$$A_{VOL} = 20 \log_{10}(V_{out}/V_{in})$$

Closed-loop voltage gain

The closed-loop voltage gain of an operational amplifier is defined as the ratio of output voltage to input voltage measured with a small proportion of the output fed back to the input (i.e. with feedback applied). The effect of providing negative feedback is to reduce the loop voltage gain to a value which is both predictable and manageable. Practical closed-loop voltage gains range from 1 to several thousand but note that high values of voltage gain may make unacceptable restrictions on bandwidth (see later).

Closed-loop voltage gain is the ratio of output voltage to input voltage when negative feedback is applied, hence:

$$A_{VCL} = V_{out}/V_{in}$$

where A_{VCL} is the closed-loop voltage gain, V_{out} and V_{in} are the output and input voltages, respectively, under closed-loop conditions. The closed-loop voltage gain is normally very much less than the open-loop voltage gain.

Example 8.1

An operational amplifier operating with negative feedback produces an output voltage of 2 V when supplied with an input of 400 μV. Determine the value of closed-loop voltage gain and express your answer in decibels.

Solution

Now

$$A_{VCL} = V_{out}/V_{in}$$

thus

$$A_{VCL} = 2\,V/400\,\mu V = 5000$$

In decibels, the value of voltage gain will be given by:

$$A_{VCL} = 20 \log_{10}(5000) = 20 \times 3.7 = 74\,dB$$

Input resistance

The input resistance of an operational amplifier is defined as the ratio of input voltage to input current expressed in ohms. It is often expedient to assume that the input of an operational amplifier is purely resistive, although this is not the case at high frequencies where shunt capacitive reactance may become significant. The input resistance of operational amplifiers is very much dependent on the semiconductor technology employed. In practice values range from about 2 MΩ for common bipolar types to over $10^{12}\,\Omega$ for FET and CMOS devices.

Input resistance is the ratio of input voltage to input current:

$$R_{in} = V_{in}/I_{in}$$

where R_{in} is the input resistance (Ω), V_{in} is the input voltage (V) and I_{in} is the input current (A). Note that we usually assume that the input of an operational amplifier is purely resistive though this may not be the case at high frequencies where shunt capacitive reactance may become significant.

Example 8.2

An operational amplifier has an input resistance of $2\,M\Omega$. Determine the input current when an input voltage of $5\,mV$ is present.

Solution

Since

$$R_{in} = V_{in}/I_{in}$$
$$I_{in} = V_{in}/R_{in} = 5\,mV/2\,M\Omega = 2.5\,nA$$

Output resistance

The output resistance of an operational amplifier is defined as the ratio of open-circuit output voltage to short-circuit output current expressed in ohms. Typical values of output resistance range from less than $10\,k\Omega$ to around $100\,k\Omega$ depending upon the configuration and amount of feedback employed.

Output resistance is the ratio of open-circuit output voltage to short-circuit output current, hence:

$$R_{out} = V_{out(OC)}/I_{out(SC)}$$

where R_{out} is the output resistance (Ω), $V_{out(OC)}$ is the open-circuit output voltage (V) and $I_{out(SC)}$ is the short-circuit output current (A).

Input offset voltage

An ideal operational amplifier would provide zero output voltage when $0\,V$ is applied to its input. In practice, due to imperfect internal balance, there may be some small voltage present at the output. The voltage that must be applied differentially to the operational amplifier input in order to make the output voltage exactly zero is known as the input offset voltage.

Figure 8.4   Offset null facility

Offset voltage may be minimized by applying relatively large amounts of negative feedback or by using the **offset null** faciliy provided by a number of operational amplifier devices. Typical values of input offset voltage range from $1\,mV$ to $15\,mV$. Where a.c., rather than d.c., coupling is employed, offset voltage is not normally a problem and can be happily ignored.

The input offset voltage is the voltage which, when applied at the input provides an output voltage of exactly zero. Similarly, the input offset current is the current which, when applied at the input, provides an output voltage of exactly zero. (Note that, due to imperfect balance and very high internal gain a small output voltage may appear with no input present.) Offset may be minimized by applying large amounts of negative feedback or by using the offset null facility provided by a number of operational amplifiers (see Fig. 8.4).

Full-power bandwidth

The full-power bandwidth is equivalent to the frequency at which the maximum undistorted peak output voltage swing falls to 0.707 of its low frequency (d.c.) value (the sinusoidal input voltage remaining constant). Typical full-power bandwidths range from $10\,kHz$ to over $1\,MHz$ for some high-speed devices.

Slew rate

Slew rate is the rate of change of output voltage with time, when a rectangular step input voltage is applied. Slew rate is measured in V/s (or V/µs) and typical values range from $0.2\,V/µs$ to over $20\,V/µs$. Slew rate imposes a imitation on circuits in which large amplitude pulses rather than small amplitude sinusoidal signals are likely to be encountered.

The slew-rate of an operational amplifier is the rate of change of output voltage with time in response to a perfect step-function input. Hence:

$$\text{slew-rate} = \mathrm{d}V_{\text{out}}/\mathrm{d}t$$

where $\mathrm{d}V_{\text{out}}$ is the change in output voltage (V) and $\mathrm{d}t$ is the corresponding interval of time (s).

Common-mode rejection ratio

Common-mode rejection ratio is a measure of an operational amplifier's ability to ignore signals simultaneously present on both inputs (i.e. 'common-mode' signals) in preference to signals applied differentially. Common-mode rejection ratio is defined as the ratio of differential voltage gain to common-mode voltage gain.

Common-mode rejection ratio is usually specified in decibels (see Appendix 5) and typical values range from 80 dB to 110 dB. Common-mode rejection ratio (CMRR) is the ratio of differential voltage gain to common-mode voltage gain (usually expressed in dB). Hence:

$$\text{CMRR} = 20\log_{10}\frac{(A_{\text{VOL(DM)}})}{(A_{\text{VOL(CM)}})}$$

where $A_{\text{VOL(DM)}}$ is the open-loop voltage gain in differential mode (equal to A_{VOL}) and $A_{\text{VOL(CM)}}$ is the open-loop voltage gain in common mode (i.e. signal applied with both inputs connected together). CMRR is thus a measure of an operational amplifier's ability to reject signals (e.g. noise) which are simultaneously present on both inputs.

Example 8.3

With no feedback applied, an operational amplifier produces an output of 10 V from a differential input of 50 μV. With the inputs shorted together, the same operational amplifier produces an output of 500 mV when an input of 2 V is present. Determine the value of common-mode rejection ratio.

Solution

First we need to find the values of differential and common-mode voltage gain ($A_{\text{VOL(DM)}}$ and $A_{\text{VOL(CM)}}$):

$$A_{\text{VOL(DM)}} = V_{\text{out(DM)}}/V_{\text{in(DM)}} = 10\,\text{V}/50\,\mu\text{V}$$
$$= 200\,000$$

and

$$A_{\text{VOL(CM)}} = V_{\text{out(CM)}}/V_{\text{in(CM)}} = 500\,\text{mV}/2\,\text{V} = 0.25$$

Now

$$\text{CMRR} = 20\log_{10}\frac{(A_{\text{VOL(DM)}})}{(A_{\text{VOL(CM)}})}$$
$$= 20\log_{10}(200\,000/0.25)$$

thus

$$\text{CMRR} = 20 \times \log_{10}(800\,000) = 118\,\text{dB}$$

Maximum output voltage swing

The maximum output voltage swing produced by an operational amplifier is the maximum range of output voltages that the device can produce without distortion. Normally these will be symmetrical about 0 V and within a volt or so of the supply voltage rails (both positive and negative).

Operational amplifier characteristics

Having defined the terminology applied to operational amplifiers we shall now consider the characteristics of an 'ideal' operational amplifier. The desirable characteristics for an operational amplifier are summarized below:

(a) The open-loop voltage gain should be very high (ideally infinite).
(b) The input resistance should be very high (ideally infinite).
(c) The output resistance should be very low (ideally zero).
(d) Full-power bandwidth should be as wide as possible.
(e) Slew-rate should be as large as possible.
(f) Input offset should be as small as possible.
(g) Common-mode rejection ratio should be as large as possible.

The characteristics of modern IC operational amplifiers come very close to those of an 'ideal' operational amplifier (see Table 8.1).

Gain and bandwidth

It is important to note that, since the product of gain and bandwidth is a constant for any particular operational amplifier, an increase in gain can only

Table 8.1 Characteristics of ideal and real operational amplifiers

Parameter	Ideal	Real
Voltage gain	Infinite	100 000
Input resistance	Infinite	100 MΩ
Output resistance	Zero	20 Ω
Bandwidth	Infinite	2 MHz

Table 8.2 Table showing the relationship between voltage gain and bandwidth for Fig. 8.5

Voltage gain (A_V)	Bandwidth
1	d.c. to 1 MHz
10	d.c. to 100 kHz
100	d.c. to 10 kHz
1 000	d.c. to 1 kHz
10 000	d.c. to 100 Hz
100 000	d.c. to 10 Hz

Figure 8.5 Gain plotted against bandwidth for a typical operational amplifier

be achieved at the expense of bandwidth, and vice versa.

Figure 8.5 shows the relationship between voltage gain and bandwidth for a typical operational amplifier (note that axes use logarithmic rather than linear scales). The open-loop voltage gain (i.e. that obtained with no feedback applied) is 100 000 (or 100 dB) and the bandwidth obtained in this condition is a mere 10 Hz. The effect of applying increasing amounts of negative feedback (and consequently reducing the gain to a more manageable amount) is that the bandwidth increases in direct proportion.

Frequency response curves have been added to Fig. 8.5 to show the effect on the bandwidth of making the closed-loop gains equal to 10 000, 1000,

100, and 10. Table 8.2 summarizes these results. You should also note that the (gain × bandwidth) product for this amplifier is 1×10^6 Hz (i.e. 1 MHz).

Example 8.4

The open-loop frequency response of an operational amplifier is shown in Fig. 8.6. Determine the bandwidth of the amplifier if the closed-loop voltage gain is set at 46 dB.

Solution

We can determine the bandwidth of the amplifier when the closed-loop voltage gain is set to 46 dB

Voltage gain (dB)

46dB
(A_v=200)

Frequency (Hz)

10kHz

1M

Gain x bandwidth
product = 2 x 10⁶ Hz

Figure 8.6

Figure 8.7   Inverting amplifier showing currents and voltages

by constructing a line and noting the intercept point on the response curve (as shown in Fig. 8.6). This shows that the bandwidth will be 10 kHz (note that, for this operational amplifier, the (gain × bandwidth) product is 2×10^6 Hz (or 2 MHz).

Inverting operational amplifier stage

Figure 8.7 shows the circuit of an operational amplifier with negative feedback applied. For the sake of our explanation we will assume that the operational amplifier is 'ideal'. Now consider what happens when a small positive input voltage is applied. This voltage (V_{in}) produces a current (I_{in}) flowing in the input resistor in the direction shown in Fig. 8.7.

Since the operational amplifier is 'ideal' we will assume that:

(a) the input resistance (i.e. the resistance that appears between the inverting and non-inverting input terminals, R_{ic}) is infinite;

(b) the open-loop voltage gain (i.e. the ratio of V_{out} to V_{in} with no feedback applied) is infinite.

As a consequence of (a) and (b):

(i) the voltage appearing between the inverting and non-inverting inputs (V_{ic}) will be zero; and

(ii) the current flowing into the chip (I_{ic}) will be zero (recall that $I_{ic} = V_{ic}/R_{ic}$ and R_{ic} is infinite).

Applying Kirchhoff's current law at node A gives:

$$I_{in} = I_{ic} + I_f \text{ but } I_{ic} = 0 \text{ thus } I_{in} = I_f \qquad (1)$$

(this shows that the current in the feedback resistor, $R2$, is the same as the input current, I_{in}).

Applying Kirchhoff's voltage law to loop A gives:

$$V_{in} = (I_{in} \times R1) + V_{ic} \text{ but } V_{ic} = 0$$
$$\text{thus } V_{in} = I_{in} \times R1 \qquad (2)$$

Applying Kirchhoff's voltage law to loop B gives:

$$V_{out} = -V_{ic} + (I_f \times R2) \text{ but } V_{ic} = 0$$

$$\text{thus } V_{out} = I_f \times R2 \tag{3}$$

Combining (1) and (3) gives:

$$V_{out} = I_{in} \times R2 \tag{4}$$

The voltage gain of the stage is given by:

$$A_V = \frac{V_{out}}{V_{in}} \tag{5}$$

Combining (4) and (2) with (5) gives:

$$A_V = \frac{I_{in} \times R2}{I_{in} \times R1} = \frac{R2}{R1}$$

Improving symmetry

To preserve symmetry and minimize offset voltage, a third resistor is often included in series with the non-inverting input (see Fig. 8.8). The value of this resistor should be equivalent to the parallel combination of $R1$ and $R2$. Hence:

$$R3 = \frac{R1 \times R2}{R1 + R2}$$

Operational amplifier circuit configurations

The three basic configurations for operational voltage amplifiers are shown in Figs 8.9, 8.10 and 8.11. Supply rails have been omitted from these diagrams for clarity but are assumed to be

Figure 8.8   Improving the symmetry of an inverting amplifier

Figure 8.9   Basic inverting amplifier circuit

Figure 8.10   Basic non-inverting amplifier circuit

$$R3 = R1$$
$$R4 = R2$$

Figure 8.11   Basic differential amplifier circuit

symmetrical about 0 V, as in Fig. 8.3. All three of these basic arrangements are d.c. coupled and their characteristics are summarized in Table 8.3.

Tailoring the frequency response

All of the amplifier circuits described previously have used direct coupling and thus have frequency response characteristics which extend to d.c. This,

Table 8.3   Characteristics of the operational amplifier circuits shown in Figs 8.9, 8.10 and 8.11

Amplifier type	Input resistance	Voltage gain	Phase shift
Inverting amplifier (Fig. 8.9)	$R1$	$R2/R1$	$180°$
Non-inverting amplifier (Fig. 8.10)	$R_{\text{in}} \times \dfrac{A_{\text{OL}}^*}{1 + (R2/R1)}$	$1 + (R2/R1)$	$0°$
Differential amplifier (Fig. 8.11)	$2R1$	$R2/R1$	$180°$

*Where R_{in} is the input resistance of the operational amplifier, and A_{OL} is the open-loop voltage gain of the operational amplifier.

of course, is undesirable for many applications, particularly where a wanted a.c. signal may be superimposed on an unwanted d.c. voltage level. In such cases a capacitor of appropriate value may be inserted in series with the input, as shown in Fig. 8.12. The value of this capacitor should be chosen so that its reactance is very much smaller than the input resistance at the lower applied input frequency. The effect of the capacitor on an amplifier's frequency response is shown in Fig. 8.13.

We can also use a capacitor to restrict the upper frequency response of an amplifier. This time, the capacitor is connected as part of the feedback path. Indeed, by selecting appropriate values of capacitor, the frequency response of an inverting operational voltage amplifier may be very easily tailored to suit individual requirements (see Figs 8.14 and 8.15). The lower cut-off frequency is determined by the value of the input capacitance, $C1$, and input resistance, $R1$. The lower cut-off frequency is given by:

$$f2 = \frac{1}{2\pi C1 R1} = \frac{0.159}{C1 R1} \text{ Hz}$$

where $C1$ is in farads and $R1$ is in ohms.

Provided the upper frequency response is not limited by the gain × bandwidth product, the upper cut-off frequency will be determined by the feedback capacitance, $C2$, and feedback resistance, $R2$, such that:

$$f2 = \frac{2}{2\pi C2 R2} = \frac{0.159}{C2 R2} \text{ Hz}$$

where $C2$ is in farads and $R2$ is in ohms.

Figure 8.12   Inverting amplifier with a.c. input coupling

Figure 8.13   Effect of $C1$ on the frequency response of the circuit shown in Fig. 8.12

Figure 8.14 Amplifier with tailored frequency response

Example 8.5

An audio voltage amplifier is to operate according to the following specification:

Voltage gain	100
Input resistance (at mid-band)	10 kΩ
Phase shift (at mid-band)	180°
Lower cut-off frequency	250 Hz
Upper cut-off frequency	15 kHz

Devise a circuit to satisfy the above specification using an operational amplifier.

Solution

To make things a little easier, we can break the problem down into manageable parts.

Since 180° phase shift is required, we shall base our circuit on a single operational amplifier with capacitors to define the upper and lower cut-off frequencies, as shown in Fig. 8.15.

The nominal input resistance is the same as the value for $R1$. Thus:

$$R1 = 10 \,\text{k}\Omega$$

To determine the value of $R2$ we can make use of the formula for mid-band voltage gain:

$$A_V = R2/R1$$

thus

$$R2 = A_V \times R1 = 100 \times 10 \,\text{k}\Omega = 100 \,\text{k}\Omega$$

To determine the value of $C1$ we will use the formula for the low frequency cut-off:

Figure 8.15 Frequency response of the circuit shown in Fig. 8.14

$$f1 = \frac{1}{2\pi C1 R1} = \frac{0.159}{C1 R1}\ \text{Hz}$$

thus

$$C1 = \frac{0.159}{f1 R1} = \frac{0.159}{250 \times 10\,000}$$
$$= 63.6 \times 10^{-9}\,\text{F} = 63.6\,\text{nF}$$

Finally, to determine the value of $C2$ we will use the formula for high frequency cut-off:

$$f2 = \frac{1}{2\pi C2 R2} = \frac{0.159}{C2 R2}\ \text{Hz}$$

thus

$$C2 = \frac{0.159}{f2 R2} = \frac{0.159}{15\,000 \times 100\,000}$$
$$= 0.106 \times 10^{-9}\,\text{F} = 106\,\text{pF}$$

Operational amplifier types and packages

Operational amplifiers are available packaged singly, in pairs (dual types), or in fours (quad types). The majority of operational amplifiers are supplied in dual-in-line (DIL) packages but TO5 packages are also used. The 081, for example, is a single general-purpose BIFET operational amplifier housed in an 8-pin DIL package. This device is also available in dual (082) and quad (084) forms (see Fig. 8.16). Table 8.4 summarizes the data for some common operational amplifier types.

Important formulae introduced in this chapter

Open-loop voltage gain:
(page 139)

$$A_{\text{VOL}} = V_{\text{out}}/V_{\text{in}}$$
$$A_{\text{VOL}} = 20 \log_{10} (V_{\text{out}}/V_{\text{in}})$$

Input resistance:
(page 140)

$$R_{\text{in}} = V_{\text{in}}/I_{\text{in}}$$

Output resistance:
(page 140)

$$R_{\text{out}} = V_{\text{out(OC)}}/I_{\text{out(SC)}}$$

Figure 8.16　Single, dual and quad integrated circuit DIL packages

Slew-rate:
(page 141)

$$S = \mathrm{d}V_{\text{out}}/\mathrm{d}t$$

Common-mode rejection ratio:
(page 141)

$$\text{CMRR} = 20 \log_{10} \frac{(A_{\text{VOL(DM)}})}{(A_{\text{VOL(CM)}})}$$

Voltage gain of an inverting amplifier:
(page 144)

$$A_{\text{V}} = \frac{R_2}{R_1}$$

Table 8.4 Characteristics of some common operational amplifiers

Device	Type	Supply voltage range (V)	Open-loop voltage gain (dB)	Input bias current	Slew rate (V/μs)
AD548	Bipolar	4.5 to 18	100 min.	0.01 nA	1.8
AD711	FET	4.5 to 18	100	25 pA	20
CA3140	CMOS	4 to 36 (or ±2 to ±18)	100	5 pA	9
LF347	FET	5 to 18	110	50 pA	13
LM301	Bipolar	5 to 18	88	70 nA	0.4
LM348	Bipolar	10 to 18	96	30 nA	0.6
TL071	FET	3 to 18	106	30 pA	13
741	Bipolar	5 to 18	106	80 nA	0.5

Voltage gain of a non-inverting amplifier:
(page 145)

$$A_V = 1 + \frac{R_2}{R_1}$$

Upper cut-off frequency:
(page 147)

$$f2 = \frac{1}{2\pi C_2 R_2}$$

Lower cut-off frequency:
(page 147)

$$f1 = \frac{1}{2\pi C_1 R_1}$$

Symbols introduced in this chapter

Operational amplifier

Figure 8.17

Problems

8.1 An operational amplifier has an open-loop voltage gain of 100 dB. If the inverting input is held at 0 V and the non-inverting input is connected to a voltage source of 0.1 mV, determine the output voltage.

8.2 The open-loop frequency response of an operational amplifier is shown in Fig. 8.18. Determine:

(a) the voltage gain for a bandwidth of 60 kHz;
(b) the bandwidth for a voltage gain of 20 dB.

8.3 An inverting operational amplifier is required to have a voltage gain of 40 and an input resistance of 5 kΩ. Determine the value of feedback resistance required.

8.4 An operational amplifier is connected in differential mode in the circuit shown in Fig. 8.19. If $R2 = R4 = 20$ kΩ and $R1 = R3 = 1$ kΩ, determine the output voltage when the following values of input voltages (measured with respect to common) are applied:

		V1	V2
(a)		0 V	+50 mV
(b)		+50 mV	0 V
(c)		+50 mV	+50 mV
(d)		+50 mV	−50 mV
(e)		−50 mV	+50 mV
(f)		−50 mV	−50 mV

8.5 An operational amplifier has a gain × bandwidth product of 2×10^5. Estimate the band-

Voltage gain (dB)

Figure 8.18

Figure 8.19

Figure 8.21

Figure 8.20

width when the device is configured for closed-loop voltage gains of:

(a) 50; and
(b) 2000.

8.6 Determine the maximum and minimum voltage gain produced by the circuit shown in Fig. 8.20.

8.7 Determine mid-band voltage gain and upper and lower cut-off frequencies for the amplifier stage shown in Fig. 8.21.

8.8   An audio voltage amplifier is to operate according to the following specification:

Voltage gain	50
Input resistance (at mid-band)	$5\,k\Omega$
Mid-band phase shift:	$180°$

Lower cut-off frequency	$10\,Hz$
Upper cut-off frequency	$20\,kHz$

Devise a circuit to satisfy the above specification using an operational amplifier.

(Answers to these problems will be found on page 261.)

9

Oscillators

This chapter describes circuits that generate sine wave, square wave, and triangular waveforms. These oscillator circuits form the basis of clocks and timing arrangements as well as signal and function generators.

In Chapter 7, we showed how negative feedback can be applied to an amplifier to form the basis of a stage which has a precisely controlled gain. An alternative form of feedback, where the output is fed back in such a way as to reinforce the input (rather than to subtract from it), is known as **positive feedback**.

Figure 9.1 shows the block diagram of an amplifier stage with positive feedback applied. Note that the amplifier provides a phase shift of 180° and the feedback network provides a further 180°. Thus the overall phase shift is 0°. The overall voltage gain is given by:

$$\text{overall gain} = \frac{V_{\text{out}}}{V_{\text{in}}}$$

Now

$V'_{\text{in}} = V_{\text{in}} + \beta V_{\text{out}}$ (by applying Kirchhoff's voltage law)

thus

$$V_{\text{in}} = V'_{\text{in}} - \beta V_{\text{out}}$$

and

$V_{\text{out}} = A_{\text{v}} \times V'_{\text{in}}$ (A_{v} is the **internal** gain of the amplifier)

Hence,

$$\text{overall gain} = \frac{A_{\text{v}} \times V'_{\text{in}}}{V'_{\text{in}} - \beta V_{\text{out}}} = \frac{A_{\text{v}} \times V'_{\text{in}}}{V'_{\text{in}} - \beta(A_{\text{v}} \times V'_{\text{in}})}$$

Thus

$$\text{overall gain} = \frac{A_{\text{v}}}{1 - \beta A_{\text{v}}}$$

Now consider what will happen when the **loop gain** (βA_{v}) approaches unity. The denominator ($1 - \beta A_{\text{v}}$) will become close to zero. This will have the effect of increasing the overall gain, i.e. the overall gain with positive feedback applied will be greater than the gain without feedback.

It is worth illustrating this difficult concept using some practical figures. Assume that you

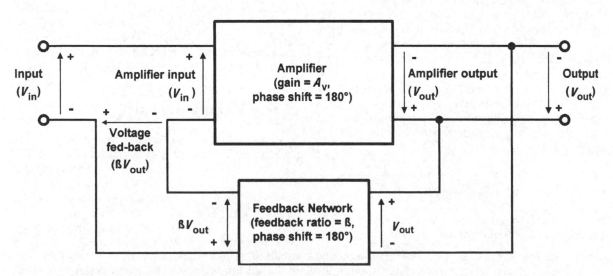

Figure 9.1   Amplifier with positive feedback applied

have an amplifier with a gain of 9 and one-tenth of the output is fed back to the input (i.e. $\beta = 0.1$). In this case the loop gain ($\beta \times A_V$) is 0.9.

With negative feedback applied (see Chapter 7) the overall voltage gain will fall to:

$$\frac{A_V}{1 + \beta A_V} = \frac{9}{1 + 0.1 \times 9} = \frac{9}{1 + 0.9} = \frac{9}{1.9} = 4.7$$

With positive feedback applied the overall voltage gain will be:

$$\frac{A_V}{1 - \beta A_V} = \frac{9}{1 - 0.1 \times 9} = \frac{9}{1 - 0.9} = \frac{9}{0.1} = 90$$

Now assume that you have an amplifier with a gain of 10 and, once again, one-tenth of the output is fed back to the input (i.e. $\beta = 0.1$). In this example the loop gain ($\beta \times A_V$) is exactly 1.

With negative feedback applied (see Chapter 7) the overall voltage gain will fall to:

$$\frac{A_V}{1 + \beta A_V} = \frac{10}{1 + 0.1 \times 10} = \frac{10}{1 + 1} = \frac{10}{2} = 5$$

With positive feedback applied the overall voltage gain will be:

$$\frac{A_V}{1 - \beta A_V} = \frac{10}{1 - 0.1 \times 10} = \frac{10}{1 - 1} = \frac{10}{0} = \text{infinity}$$

This simple example shows that a loop gain of unity (or larger) will result in infinite gain and an amplifier which is unstable. In fact, the amplifier will **oscillate** since any disturbance will be amplified and result in an output. Clearly, as far as an amplifier is concerned, positive feedback may have an undesirable effect – instead of reducing the overall gain the effect is that of reinforcing any signal present and the output can build up into continuous oscillation if the loop gain is 1 or greater. To put this another way, oscillator circuits can simply be thought of as amplifiers that generate an output signal without the need for an input!

Conditions for oscillation

From the foregoing we can deduce that the conditions for oscillation are:

(a) the feedback must be positive (i.e. the signal fed back must arrive back in-phase with the signal at the input);

Figure 9.2   Sine wave oscillator based on a three-stage *C–R* ladder network

(b) the overall loop voltage gain must be greater than 1 (i.e. the amplifier's gain must be sufficient to overcome the losses associated with any frequency selective feedback network).

Hence, to create an oscillator we simply need an amplifier with sufficient gain to overcome the losses of the network that provide positive feedback. Assuming that the amplifier provides 180° phase shift, the frequency of oscillation will be that at which there is 180° phase shift in the feedback network. A number of circuits can be used to provide 180° phase shift, one of the simplest being a three-stage *C–R* ladder network. Alternatively, if the amplifier produces 0° phase shift, the circuit will oscillate at the frequency at which the feedback network produces 0° phase shift. In both cases, the essential point is that the feedback should be positive so that the output signal arrives back at the input in such a sense as to reinforce the original signal.

Ladder network oscillator

A simple phase-shift oscillator based on a three-stage *C–R* ladder network is shown in Fig. 9.2. TR1 operates as a conventional common-emitter amplifier stage with *R*1 and *R*2 providing base bias potential and *R*3 and *C*1 providing emitter stabilization. The total phase shift provided by the *C–R* ladder network (connected between collector and base) is 180° at the frequency of oscillation. The transistor provides the other 180° phase shift in order to realize an overall phase shift of 360° or 0°.

The frequency of oscillation of the circuit shown in Fig. 9.2 is given by:

$$f = \frac{1}{2\pi\sqrt{6}CR}$$

The loss associated with the ladder network is 29, thus the amplifier must provide a gain of at least 29 in order for the circuit to oscillate.

Example 9.1

Determine the frequency of oscillation of a ladder network oscillator where $C = 10\,\text{nF}$ and $R = 10\,\text{k}\Omega$.

Solution

Using $f = 1/2\pi\sqrt{6}CR$ gives

$$f = \frac{1}{6.28 \times 2.45 \times 10 \times 10^{-9} \times 10 \times 10^3}\ \text{Hz}$$

thus

$$f = \frac{1}{6.28 \times 2.45 \times 10^{-4}} = \frac{10^4}{15.386} = 647\,\text{Hz}$$

Wien bridge oscillator

An alternative approach to providing the phase shift required is the use of a Wien bridge network (Fig. 9.3). Like the C–R ladder, this network provides a phase shift which varies with frequency. The input signal is applied to A and B while the output is taken from C and D. At one particular frequency, the phase shift produced by the network will be exactly zero (i.e. the input and output signals will be in-phase). If we connect the network to an amplifier producing 0° phase shift which has sufficient gain to overcome the losses of the Wien bridge, oscillation will result.

The minimum amplifier gain required to sustain oscillation is given by:

$$A_V = 1 + C1/C2 + R2/R1$$

and the frequency at which the phase shift will be zero is given by:

$$f = \frac{1}{2\pi\sqrt{(C1C2R1C2)}}$$

In practice, we normally make $R1 = R2$ and $C1 = C2$ hence:

$$A_V = 1 + C/C + R/R = 1 - 1 + 1 = 3$$

Figure 9.3 A Wien bridge network

and the frequency at which the phase shift will be zero is given by:

$$f = \frac{1}{2\pi\sqrt{(RRCC)}} = \frac{1}{2\pi CR}$$

where $R = R1 = R2$ and $C = C1 = C2$.

Example 9.2

Figure 9.4 shows the circuit of a Wien bridge oscillator based on an operational amplifier. If $C1 = C2 = 100\,\text{nF}$, determine the output frequencies produced by this arrangement (a) when $R1 = R2 = 1\,\text{k}\Omega$ and (b) when $R1 = R2 = 6\,\text{k}\Omega$.

Solution

(a) When $R1 = R2 = 1\,\text{k}\Omega$

$$f = \frac{1}{2\pi CR}$$

Where $R = R1 = R2$ and $C = C1 = C2$.

Thus

$$f = \frac{1}{6.28 \times 100 \times 10^{-9} \times 1 \times 10^3}$$

$$= \frac{1}{6.28 \times 10^{-4}}$$

Figure 9.4   Sine wave oscillator based on a Wien bridge network (see Example 3.2)

or

$$f = \frac{10\,000}{6.28} = 1592\,\text{Hz} = 1.592\,\text{kHz}$$

(b) When $R1 = R2 = 6\,\text{k}\Omega$

$$f = \frac{1}{2\pi CR} = \frac{1}{6.28 \times 100 \times 10^{-9} \times 6 \times 10^{3}}$$

$$= \frac{1}{37.68 \times 10^{-4}}$$

or

$$f = \frac{10\,000}{37.68} = 265.4\,\text{Hz}$$

Multivibrators

There are many occasions when we require a square wave output from an oscillator rather than a sine wave output. Multivibrators are a family of oscillator circuits that produce output waveforms consisting of one or more rectangular pulses. The term 'multivibrator' simply originates from the fact that this type of waveform is rich in harmonics (i.e. 'multiple vibrations').

Multivibrators use regenerative (positive) feedback; the active devices present within the oscillator circuit being operated as switches, being alternately cut off and driven into saturation.

The principal types of multivibrator are:

(a) **astable multivibrators** that provide a continuous train of pulses (these are sometimes also referred to as **free-running multivibrators**);
(b) **monostable multivibrators** that produce a single output pulse (they have one stable state and are thus sometimes also referred to as **one-shot** circuits);
(c) **bistable multivibrators** that have two stable states and require a trigger pulse or control signal to change from one state to another.

The astable multivibrator

Figure 9.5 shows a classic form of astable multivibrator based on two transistors. Figure 9.6 shows how this circuit can be redrawn in an arrangement that more closely resembles a two-stage common-emitter amplifier with its output connected back to its input.

In Fig. 9.5, the values of the base resistors, $R3$ and $R4$, are such that the sufficient base current will be available to completely saturate the respective transistor. The values of the collector load resistors, $R1$ and $R2$, are very much smaller than $R3$ and $R4$. When power is first applied to the circuit, assume that TR2 saturates before TR1 when the power is first applied (in practice one transistor would always saturate before the other due to variations in component tolerances and transistor parameters).

Figure 9.5 Astable multivibrator using bipolar transistors

Figure 9.6 Circuit of Figure 9.5 redrawn to show the two common-emitter stages

As TR2 saturates, its collector voltage will fall rapidly from $+V_{CC}$ to 0 V. This drop in voltage will be transferred to the base of TR1 via C2. This negative going voltage will ensure that TR1 is initially placed in the non-conducting state. As long as TR1 remains cut off, TR2 will continue to be saturated. During this time, C2 will charge via R4 and TR1's base voltage will rise exponentially from $-V_{CC}$ towards $+V_{CC}$. However, TR1's base voltage will not rise much above 0 V because, as soon as it reaches $+0.7$ V (sufficient to cause base current to flow) TR1 will begin to conduct. As TR1 begins to turn on, its collector voltage will rapidly fall from $+V_{CC}$ to 0 V. This fall in voltage is transferred to the base of TR2 via C1 and, as a consequence, TR2 will turn off. C1 will then charge via R3 and TR2's base voltage will rise exponentially from $-V_{CC}$ towards $+V_{CC}$. As before, TR2's base voltage will not rise much above 0 V because, as soon as it reaches $+0.7$ V (sufficient to cause base current to flow), TR2 will start to conduct. The cycle is then repeated indefinitely.

The time for which the collector voltage of TR2 is low and TR1 is high ($T1$) will be determined by the time constant, $R4 \times C2$. Similarly, the time for which the collector voltage of TR1 is low and TR2 is high ($T2$) will be determined by the time constant, $R3 \times C1$.

The following approximate relationships apply:

$$T1 = 0.7C2R4 \quad \text{and} \quad T2 = 0.7C1R3$$

Since one complete cycle of the output occurs in a time, $T = T1 + T2$, the periodic time of the output is given by:

$$T = 0.7 (C2R4 + C1R3)$$

Finally, we often require a symmetrical 'square wave' output where $T1 = T2$. To obtain such an output, we should make $R3 = R4$ and $C1 = C2$, in which case the periodic time of the output will be given by:

$$T = 1.4CR$$

where $C = C1 = C2$ and $R = R3 = R4$.

Waveforms for the astable oscillator are shown in Fig. 9.7.

Example 9.3

The astable multivibrator in Fig. 9.4 is required to produce a square wave output at 1 kHz. Determine suitable values for R3 and R4 if C1 and C2 are both 10 nF.

Solution

Since a square wave is required and C1 and C2 have identical values, R3 must be made equal to R4. Now:

Figure 9.7 Waveforms for the transistor astable multivibrator

$$T = \frac{1}{f} = \frac{1}{1 \times 10^3} = 1 \times 10^{-3} \, \text{s}$$

Re-arranging $T = 1.4CR$ to make R the subject gives:

$$R = \frac{T}{1.4C} = \frac{1 \times 10^{-3}}{1.4 \times 10 \times 10^{-9}} = \frac{1 \times 10^6}{14}$$

$$= 0.071 \times 10^6 \, \Omega \, \text{ or } 71.4 \, \text{k}\Omega$$

Other forms of astable oscillator

Figure 9.8 shows the circuit diagram of an alternative form of astable oscillator which produces a triangular output waveform. Operational amplifier IC1 forms an integrating stage while IC2 is connected with positive feedback to ensure that oscillation takes place.

Assume that the output from IC2 is initially at, or near, $+V_{CC}$ and capacitor, C, is uncharged. The voltage at the output of IC2 will be passed, via R, to IC1. Capacitor, C, will start to charge and the output voltage of IC1 will begin to fall. Eventually, the output voltage will have fallen to a value that causes the polarity of the voltage at the non-inverting input of IC2 to change from positive to negative. At this point, the output of IC2 will rapidly fall to $-V_{CC}$. Again, this voltage will be passed, via R, to IC1. Capacitor, C, will then start to charge in the other direction and the output voltage of IC1 will begin to rise. Eventually, the output voltage will have risen to a value that causes the polarity of the non-inverting input of

IC2 to revert to its original (positive) state and the cycle will continue indefinitely.

The upper threshold voltage (i.e. the maximum positive value for V_{out}) will be given by:

$$V_{UT} = V_{CC} \times (R1/R2)$$

The lower threshold voltage (i.e. the maximum negative value for V_{out}) will be given by:

$$V_{LT} = -V_{CC} \times (R1/R2)$$

Single-stage astable oscillator

A simple form of astable oscillator producing a square wave output can be built using just one operational amplifier, as shown in Fig. 9.9. This circuit employs positive feedback with the output fed back to the non-inverting input via the potential divider formed by $R1$ and $R2$.

Assume that C is initially uncharged and the voltage at the inverting input is slightly less than the voltage at the non-inverting input. The output voltage will rise rapidly to $+V_{CC}$ and the voltage at the inverting input will begin to rise exponentially as capacitor C charges through R. Eventually, the voltage at the inverting input will have reached a value that causes the voltage at the inverting input to exceed that present at the non-inverting input. At this point, the output voltage will rapidly fall to $-V_{CC}$. Capacitor, C, will then start to charge in the other direction and the voltage at the inverting input will begin to fall exponentially. Eventually, the voltage at the inverting input will have reached a value that causes the voltage at the inverting input to be less than that present at the non-

Figure 9.8 Astable oscillator using operational amplifiers

Figure 9.9 Single-stage astable oscillator using an operational amplifier

Figure 9.10 Simple crystal oscillator

inverting input. At this point, the output voltage will rise rapidly to $+V_{CC}$ once again and the cycle will continue indefinitely.

The upper threshold voltage (i.e. the maximum positive value for the voltage at the inverting input) will be given by:

$$V_{UT} = V_{CC} \times \frac{R2}{R1 + R2}$$

The lower threshold voltage (i.e. the maximum negative value for the voltage at the inverting input) will be given by:

$$V_{LT} = -V_{CC} \times \frac{R2}{R1 + R2}$$

Crystal controlled oscillators

A requirement of some oscillators is that they accurately maintain an exact frequency of oscillation. In such cases, a quartz crystal can be used as the frequency determining element. The quartz crystal (a thin slice of quartz in a hermetically sealed enclosure) vibrates whenever a potential difference is applied across its faces (this phenomenon is known as the **piezoelectric effect**). The frequency of oscillation is determined by the crystal's 'cut' and physical size.

Most quartz crystals can be expected to stabilize the frequency of oscillation of a circuit to within a few parts in a million. Crystals can be manufactured for operation in fundamental mode over a frequency range extending from 100 kHz to around 20 MHz and for overtone operation from 20 MHz to well over 100 MHz. Figure 9.10 shows a simple crystal oscillator circuit in which the crystal provides feedback from the drain to the source of a junction gate FET.

Practical oscillator circuits

Figure 9.11 shows a practical sine wave oscillator based on a three-stage $C-R$ ladder network. The circuit provides an output of approximately 1 V pk–pk at 1.97 kHz.

Figure 9.11 A practical sine wave oscillator based on a phase shift ladder network

Figure 9.12 A practical sine wave oscillator based on a Wien bridge

Figure 9.13 A practical square wave oscillator based on an astable multivibrator

A practical Wien bridge oscillator is shown in Fig. 9.12. This circuit produces a sine wave output at 16 Hz. The output frequency can easily be varied by making R1 and R2 a 10 kΩ dual-gang potentiometer and connecting a fixed resistor of 680 Ω in series with each. In order to adjust the loop gain for an optimum sine wave output it may be necessary to make R3/R4 adjustable. One way of doing this is to replace both components with a 10 kΩ multi-turn potentiometer with the sliding contact taken to the inverting input of IC1.

An astable multivibrator is shown in Fig. 9.13. This circuit produces a square wave output of 5 V pk–pk at approximately 690 Hz.

A triangle wave generator is shown in Fig. 9.14. This circuit produces a symmetrical triangular output waveform at approximately 8 Hz. If desired, a simultaneous square wave output can be derived from the output of IC2. The circuit requires symmetrical supply voltage rails (not shown in Fig. 9.14) of between ±9 V and ±15 V.

Figure 9.15 shows a single-stage astable oscillator. This circuit produces a square wave output at approximately 13 Hz.

Finally, Fig. 9.16 shows a high-frequency crystal oscillator that produces an output of approximately 1 V pk–pk at 4 MHz. The precise frequency of operation depends upon the quartz crystal employed (the circuit will operate with fundamental mode crystals in the range 2 MHz to about 12 MHz).

Figure 9.14 A practical triangle wave generator

Figure 9.15 A single-stage astable oscillator which produces a square wave output

Figure 9.16 A practical high-frequency crystal oscillator

Symbol introduced in this chapter

Quartz crystal

Figure 9.17

Important formulae introduced in this chapter

Gain with positive feedback applied:
(page 151)

$$A_{\mathrm{VPFB}} = \frac{A_{\mathrm{V}}}{1 - \beta A_{\mathrm{V}}}$$

Loop gain:
(page 151)

$$A_{\mathrm{VLOOP}} = \beta A_{\mathrm{V}}$$

Output frequency of three-stage $C-R$ ladder network oscillator:
(page 153)

$$f = \frac{1}{2\pi\sqrt{6}CR}$$

Output frequency of Wien bridge oscillator:
(page 153)

$$f = \frac{1}{2\pi CR}$$

Time for which a multivibrator output is 'high':
(page 155)

$$T = 0.7C_{\mathrm{T}}R_{\mathrm{B}}$$

Periodic time for the output of a square wave multivibrator:
(page 155)

$$T = 1.4CR$$

Problems

9.1 An amplifier with a gain of 8 has 10% of its output fed back to the input. Determine the gain of the stage (a) with negative feedback, (b) with positive feedback.

9.2 A phase-shift oscillator is to operate with an output at 1 kHz. If the oscillator is based on a three-stage ladder network, determine the required values of resistance if three capacitors of 10 nF are to be used.

9.3 A Wien bridge oscillator is based on the circuit shown in Fig. 9.3 but R1 and R2 are replaced by a dual-gang potentiometer. If

Figure 9.18

$C1 = C2 = 22\,\text{nF}$ determine the values of $R1$ and $R2$ required to produce an output at exactly 400 Hz.

9.4   Determine the peak–peak voltage developed across C1 in Fig. 9.18.

9.5   An astable multivibrator circuit is required to produce an asymmetrical rectangular output which has a period of 4 ms and is to be 'high' for 1 ms and 'low' for 3 ms. If the timing capacitors are both to be 100 nF, determine the values of the two timing resistors required.

(Answers to these problems can be found on page 261.)

10

Logic circuits

This chapter introduces electronic circuits and devices that are associated with digital rather than analogue circuitry. These logic circuits are used extensively in digital systems and form the basis of clocks, counters, shift registers and timers. The chapter starts by introducing the basic logic functions (AND, OR, NAND, NOR, etc.) together with the symbols and truth tables that describe the operation of the most common logic gates. We then show how these gates can be used in simple combinational logic circuits before moving on to introduce bistable devices, counters and shift registers. The chapter concludes with a brief introduction to the two principal technologies used in modern digital logic circuits, TTL and CMOS.

Logic functions

Electronic circuits can be used to make simple decisions like:

If dark then put on the light.

and

If temperature is less then 20°C then connect the supply to the heater.

They can also be used to make more complex decisions like:

If 'hour' is greater than 11 and '24 hour clock' is not selected then display message 'pm'.

All of these logical statements are similar in form. The first two are essentially:

If {condition} *then* {action}.

while the third is a compound statement of the form:

If {condition 1} *and not* {condition 2} *then* {action}.

Both of these statements can be readily implemented using straightforward electronic circuitry. Because this circuitry is based on discrete states and since the behaviour of the circuits can be described by a set of logical statements, it is referred to as **digital logic**.

Switch and lamp logic

Consider the simple circuit shown in Fig. 10.1. In this circuit a battery is connected to a lamp via a switch. It should be obvious that the lamp will only operate when the switch is closed. There are two possible states for the switch, open and closed. We can summarize the operation of the circuit using Table 10.1.

Since the switch can only be in one of the two states (i.e. open or closed) at any given time, the open and closed conditions are mutually exclusive. Furthermore, since the switch cannot exist in any other state than completely open or completely closed (i.e. there is no intermediate or half-open state) the circuit uses binary or 'two-state' logic. We can represent the logical state of the switch using the binary digits, 0 and 1. We shall assume that a logical 0 is synonymous with open (or 'off') and logical 1 is synonymous with closed (or 'on'). Hence:

Figure 10.1 Simple switch and lamp circuit

Table 10.1 Simple switching logic

Condition	Switch	Comment
1	Open	No light produced
2	Closed	Light produced

A	LP1
0	0
1	1

Figure 10.2   Truth table for the switch and lamp

Figure 10.3   AND switch and lamp logic

Table 10.2   Simple AND switching logic

Condition	Switch A	Switch B	Comment
1	Open	Open	No light produced
2	Open	Closed	No light produced
3	Closed	Open	No light produced
4	Closed	Closed	Light produced

A	B	Y
0	0	0
0	1	0
1	0	0
1	1	1

Figure 10.4   Truth table for AND logic

Figure 10.5   OR switch and lamp logic

Switch open (off) = 0
Switch closed (on) = 1

We can now rewrite the truth table in terms of the binary states as shown in Fig. 10.2 where:

No light (off) = 0
Light (on) = 1

AND logic

Now consider the circuit with two switches shown in Fig. 10.3. Here the lamp will only operate when switch A is closed AND switch B is closed. However, let's look at the operation of the circuit in a little more detail. Since there are two switches (A and B) and there are two possible states for each switch (open or closed), there is a total of four possible conditions for the circuit. We can summarize these conditions in Table 10.2.

Since each switch can only be in one of the two states (i.e. open or closed) at any given time, the open and closed conditions are mutually exclusive. Furthermore, since the switches cannot exist in any other state than completely open or completely closed (i.e. there are no intermediate states) the circuit uses 'binary logic'. We can thus represent the logical states of the two switches by the binary digits, 0 and 1.

Once again, if we adopt the obvious convention that an open switch can be represented by 0 and a closed switch by 1, we can rewrite the truth table in terms of the binary states shown in Fig. 10.4 where

No light (off) = 0
Light (on) = 1

OR logic

Figure 10.5 shows another circuit with two switches. This circuit differs from that shown in Fig. 10.3 by virtue of the fact that the two switches are connected in parallel rather than in series. In this case the lamp will operate when either of the two switches is closed. As before, there is a total of four possible conditions for the circuit. We can summarize these conditions in Table 10.3.

Once again, adopting the convention that an open switch can be represented by 0 and a closed switch by 1, we can rewrite the truth table in terms of the binary states as shown in Fig. 10.6.

Table 10.3 Simple OR switching logic

Condition	Switch A	Switch B	Comment
1	Open	Open	No light produced
2	Open	Closed	Light produced
3	Closed	Open	Light produced
4	Closed	Closed	Light produced

A	B	Y
0	0	0
0	1	1
1	0	1
1	1	1

Figure 10.6 Truth table for OR logic

Figure 10.7

A	B	C	Y
0	0	0	0
0	0	1	0
0	1	0	0
0	1	1	0
1	0	0	0
1	0	1	1
1	1	0	1
1	1	1	1

Figure 10.8

Example 10.1

Figure 10.7 shows a simple switching circuit. Describe the logical state of switches A, B, and C in order to operate the lamp. Illustrate your answer with a truth table.

Solution

In order to operate the lamp, switch A AND either switch B OR switch C must be operated. The truth table is shown in Fig. 10.8.

Logic gates

Logic gates are circuits designed to produce the basic logic functions, AND, OR, etc. These circuits are designed to be interconnected into larger, more complex, logic circuit arrangements. Since these circuits form the basic building blocks of all digital systems, we have summarized the action of each of the gates in the next section. For each gate we have included its British Standard (BS) symbol together with its American Standard (MIL/ANSI) symbol. We have also included the truth tables and Boolean expressions (using '+' to denote OR, '.' to denote AND, and '−' to denote NOT). Note that, while inverters and buffers each have only one input, exclusive-OR gates have two inputs and the other basic gates (AND, OR, NAND and NOR) are commonly available with up to eight inputs.

Buffers (Fig. 10.9)

Buffers do not affect the logical state of a digital signal (i.e. a logic 1 input results in a logic 1 output whereas a logic 0 input results in a logic 0 output). Buffers are normally used to provide extra current drive at the output but can also be used to regularize the logic levels present at an interface. The Boolean expression for the output, Y, of a buffer with an input, X, is:

$$Y = X$$

Figure 10.9 Symbols and truth table for a buffer

Inverters (Fig. 10.10)

Inverters are used to complement the logical state (i.e. a logic 1 input results in a logic 0 output and vice versa). Inverters also provide extra current drive and, like buffers, are used in interfacing applications where they provide a means of regularizing logic levels present at the input or output of a digital system. The Boolean expression for the output, Y, of a buffer with an input, X, is:

$Y = \overline{X}$

Figure 10.10   Symbols and truth table for an inverter

AND gates (Fig. 10.11)

AND gates will only produce a logic 1 output when all inputs are simultaneously at logic 1. Any other input combination results in a logic 0 output. The Boolean expression for the output, Y, of an AND gate with inputs, A and B, is:

$Y = A.B$

Figure 10.11   Symbols and truth table for an AND gate

OR gates (Fig. 10.12)

OR gates will produce a logic 1 output whenever any one, or more, inputs are at logic 1. Putting this another way, an OR gate will only produce a logic 0 output whenever all of its inputs are simultaneously at logic 0. The Boolean expression for the output, Y, of an OR gate with inputs, A and B, is:

$Y = A + B$

Figure 10.12   Symbols and truth table for an OR gate

NAND gates (Fig. 10.13)

NAND (i.e. NOT-AND) gates will only produce a logic 0 output when all inputs are simultaneously at logic 1. Any other input combination will produce a logic 1 output. A NAND gate, therefore, is nothing more than an AND gate with its output inverted! The circle shown at the output denotes this inversion. The Boolean expression for the output, Y, of a NAND gate with inputs, A and B, is:

$Y = \overline{A.B}$

Figure 10.13   Symbols and truth table for a NAND gate

NOR gates (Fig. 10.14)

NOR (i.e. NOT-OR) gates will only produce a logic 1 output when all inputs are simultaneously

Figure 10.14   Symbols and truth table for a NOR gate

at logic 0. Any other input combination will produce a logic 0 output. A NOR gate, therefore, is simply an OR gate with its output inverted. A circle is again used to indicate inversion. The Boolean expression for the output, Y, of a NOR gate with inputs, A and B, is:

$$Y = \overline{A + B}$$

Exclusive-OR gates (Fig. 10.15)

Exclusive-OR gates will produce a logic 1 output whenever either one of the inputs is at logic 1 and the other is at logic 0. Exclusive-OR gates produce a logic 0 output whenever both inputs have the same logical state (i.e. when both are at logic 0 or both are at logic 1). The Boolean expression for the output, Y, of an exclusive-OR gate with inputs, A and B, is:

$$Y = A.\overline{B} + B.\overline{A}$$

Figure 10.15 Symbols and truth table for an exclusive-OR gate

$$Y = (B.C) + (A.C) + (A.B) + A.B.C$$

Figure 10.16

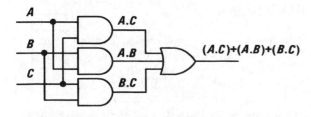

Figure 10.17

Combinational logic

By using a standard range of logic levels (i.e. voltage levels used to represent the logic 1 and logic 0 states) logic circuts can be combined together in order to solve complex logic functions.

Example 10.2

A logic circuit is to be constructed that will produce a logic 1 output whenever two, or more, of its three inputs are at logic 1.

Solution

This circuit could be more aptly referred to as a **majority vote** circuit. Its truth table is shown in Fig. 10.16. Figure 10.17 shows the logic circuitry required.

Example 10.3

Show how an arrangement of basic logic gates (AND, OR and NOT) can be used to produce the exclusive-OR function.

Solution

In order to solve this problem, consider the Boolean expression for the exclusive-OR function:

$$Y = A.\overline{B} + B.\overline{A}$$

This expression takes the form:

$$Y = P + Q \text{ where } P = A.\overline{B} \text{ and } Q = B.\overline{A}$$

$A.\overline{B}$ and $B.\overline{A}$ can be obtained using two two-input AND gates and the result (i.e. P and Q) can then be applied to a two-input OR gate.

$\overline{A}$ and $\overline{B}$ can be produced using inverters. The complete solution is shown in Fig. 10.18.

Figure 10.18

(a)

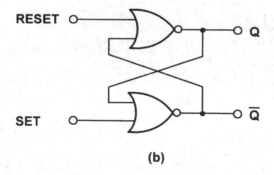

(b)

Figure 10.19 R-S bistables using cross-coupled NAND and NOR gates

Bistables

The output of a bistable has two stables states (logic 0 or logic 1) and, once **set** in one or other of these states, the device will remain at a particular logic level for an indefinite period until **reset**. A bistable thus constitutes a simple form of 'memory cell'; as it will remain in its **latched** state (whether set or reset) until a signal is applied to it in order to change its state (or until the supply is disconnected).

R-S bistables

The simplest form of bistable is the R-S bistable. This device has two inputs, SET and RESET, and complementary outputs, Q and $\bar{Q}$. A logic 1 applied to the SET input will cause the Q output to become (or remain at) logic 1 while a logic 1 applied to the RESET input will cause the Q output to become (or remain at) logic 0. In either case, the bistable will remain in its SET or RESET state until an input is applied in such a sense as to change the state.

Two simple forms of R-S bistable based on cross-coupled logic gates are shown in Fig. 10.19. Figure 10.19(a) is based on NAND gates while Fig. 10.19(b) is based on NOR gates.

The simple cross-coupled logic gate bistable has a number of serious shortcomings (consider what would happen if a logic 1 was simultaneously present on both the SET and RESET inputs!) and practical forms of bistable make use of much improved purpose-designed logic circuits such as D-type and J-K bistables.

D-type bistables

The D-type bistable has two inputs: D (standing variously for data or delay) and CLOCK (CLK). The data input (logic 0 or logic 1) is clocked into the bistable such that the output state only changes when the clock changes state. Operation is thus said to be synchronous. Additional subsidiary inputs (which are invariably active low) are provided which can be used to directly set or reset the bistable. These are usually called PRESET (PR) and CLEAR (CLR). D-type bistables are used both as latches (a simple form of memory) and as binary dividers. The simple circuit arrangement in Fig. 10.20 together with the timing diagram shown in Fig. 10.21 illustrate the operation of D-type bistables.

J-K bistables

J-K bistables have two clocked inputs (J and K), two direct inputs (PRESET and CLEAR), a CLOCK (CK) input, and outputs (Q and $\bar{Q}$). As with R-S bistables, the two outputs are complementary (i.e. when one is 0 the other is 1, and vice versa). Similarly, the PRESET and CLEAR inputs are invariably both active low (i.e. a 0 on the

Figure 10.20   D-type bistable operation

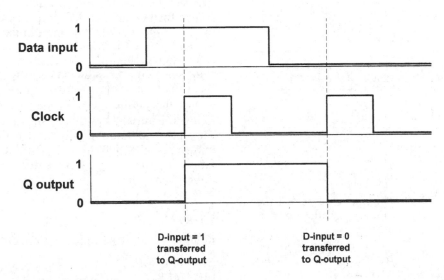

Figure 10.21   Timing diagram for the D-type bistable

PRESET input will set the Q output to 1 whereas a 0 on the CLEAR input will set the Q output to 0). Table 10.4 summarizes the operation of a J-K bistable for various input states.

J-K bistables are the most sophisticated and flexible of the bistable types and they can be configured in various ways including binary dividers, shift registers, and latches. Figure 10.22 shows the arrangement of a four-stage binary counter based on J-K bistables. The **timing diagram** for this circuit is shown in Fig. 10.23. Each stage successively divides the clock input signal by a factor of two. Note that a logic 1 input is transferred to the respective Q-output on the falling edge of the clock pulse and all J and K inputs must be taken to logic 1 to enable binary counting.

Figure 10.24 shows the arrangement of a four-stage shift register based on J-K bistables. The

timing diagram for this circuit is shown in Fig. 10.25. Note that each stage successively feeds data (Q output) to the next stage. Note that all data transfer occurs on the falling edge of the clock pulse.

Example 10.4

A certain logic arrangement is to produce the pulse train shown in Fig. 10.26. Devise a logic circuit arrangement that will generate this pulse train from a regular square wave input.

Solution

A two-stage binary divider (based on J-K bistables) can be used together with a two-input AND gate as shown in Fig. 10.27. The waveforms for this logic arrangement are shown in Fig. 10.28.

Table 10.4 Input and output states for a J-K bistable

Inputs		Output	Comments
J	K	Q_{N+1}	
0	0	Q_N	No change in state of the Q output on the next clock transition
0	1	0	Q output changes to 0 (i.e. Q is reset) on the next clock transition
1	0	1	Q output changes to 1 (i.e. Q is set) on the next clock transition
1	1	Q_N	Q output changes to the opposite state on the next clock transition

Note: Q_{N+1} means 'Q after next clock transition' while Q_N means 'Q in whatever state it was before'.

Inputs		Output	Comments
Preset	Clear	Q	
0	0	?	Indeterminate
0	1	0	Q output changes to 1 (i.e. Q is reset) regardless of the clock state
1	0	1	Q output changes to 1 (i.e. Q is set) regardless of the clock state
1	1	–	Enables clocked operation (refer to previous table for state of Q_{N+1})

Note: The preset and clear inputs operate regardless of the clock.

Integrated circuit logic devices

The task of realizing a complex logic circuit is made simple with the aid of digital integrated circuits. Such devices are classified according to the semiconductor technology used in their fabrication (the **logic family** to which a device belongs is largely instrumental in determining its operational characteristics, such as power consumption, speed, and immunity to noise).

The two basic logic families are CMOS (complementary metal oxide semiconductor) and TTL (transistor transistor logic). Each of these families is then further sub-divided. Representative circuits for a two-input AND gate in both technologies are shown in Figs 10.29 and 10.30.

The most common family of TTL logic devices is known as the 74-series. Devices from this family are coded with the prefix number 74. Variants within the family are identified by letters which follow the initial 74 prefix, as shown in Table 10.5. The most common family of CMOS devices is known as the 4000-series. Variants within the family are identified by the suffix letters given in Table 10.6

Example 10.5

Identify each of the following integrated circuits:

(i) 4001UBE;
(ii) 74LS14.

Solution

Integrated circuit (i) is an improved (unbuffered) version of the CMOS 4001 device.

Figure 10.22 Four-stage binary counter using J-K bistables

Figure 10.23   Timing diagram for the four-stage binary counter

Figure 10.24   Four-stage shift register using J-K bistables

Integrated circuit (ii) is a low-power Schottky version of the TTL 7414 device.

Date codes

It is also worth noting that the vast majority of logic devices and other digital integrated circuits are marked with a four digit date code. The first two digits of this code give the last two digits of the year of manufacture while the last two digits specify the week of manufacture. The code often appears alongside or below the device code.

Example 10.6

An integrated circuit marked '4050B 8832'. What type of device is it and when was it manufactured?

Solution

The device is a buffered CMOS 4050 manufactured in the 32nd week of 1988.

Figure 10.25   Timing diagram for the four-stage shift register

Figure 10.26

Figure 10.27

Figure 10.28

Figure 10.29   Two-input TTL NAND gate

Logic levels

Logic levels are simply the range of voltages used to represent the logic states 0 and 1. The logic levels for CMOS differ markedly from those associated with TTL. In particular, CMOS logic levels are relative to the supply voltage used while the logic levels associated with TTL devices tend to be absolute (see Table 10.7).

Noise margin

The noise margin of a logic device is a measure of its ability to reject noise; the larger the noise margin the better is its ability to perform in an environment in which noise is present. Noise margin is defined as the difference between the minimum values of high state output and high state input voltage and the maximum values of

Figure 10.30   Two-input CMOS NAND gate

Table 10.5   TTL device coding

Infix	Meaning
None	Standard TTL device
ALS	Advanced low-power Schottky
C	CMOS version of a TTL device
F	'Fast' – a high-speed version of the device
H	High-speed version
S	Schottky input configuration (improved speed and noise immunity)
HC	High-speed CMOS version (CMOS compatible inputs)
HCT	High-speed CMOS version (TTL compatible inputs)
LS	Low-power Schottky

Table 10.6   CMOS device coding The most common family of CMOS devices is known as the 4000-series. Variants within the family are identified by suffix letters as follows:

Suffix	Meaning
None	Standard CMOS device
A	Standard (unbuffered) CMOS device
B, BE	Improved (buffered) CMOS device
UB, UBE	Improved (unbuffered) CMOS device

low state output and low state input voltage. Hence:

$$\text{noise margin} = V_{\text{oh(MIN)}} - V_{\text{ih(MIN)}}$$

or

$$\text{noise margin} = V_{\text{ol(MAX)}} - V_{\text{il(MAX)}}$$

Table 10.7   CMOS and TTL characteristics

Condition	CMOS	TTL
Logic 1	More than $(2/3)\,V_{\text{DD}}$	More than 2 V
Logic 0	Less than $(1/3)\,V_{\text{DD}}$	Less than 0.8 V
Indeterminate	Between $(1/3)\,V_{\text{DD}}$ and $(2/3)\,V_{\text{DD}}$	Between 0.8 V and 2 V

Note: V_{DD} is the positive supply associated with CMOS devices.

Figure 10.31   Logic levels and noise margins for TTL and CMOS device:

where $V_{\text{oh(MIN)}}$ is the minimum value of high state (logic 1) output voltage, $V_{\text{ih(MIN)}}$ is the minimum value of high state (logic 1) input voltage, $V_{\text{ol(MAX)}}$ is the maximum value of low state (logic 0) output voltage, and $V_{\text{il(MIN)}}$ is the minimum value of low state (logic 0) input voltage.

The noise margin for standard 7400 series TTL is typically 400 mV while that for CMOS is (1/3) V_{DD}, as shown in Fig. 10.31.

Table 10.8 compares the more important characteristics of common members of the TTL family with their buffered CMOS logic counter-parts. Finally, Fig. 10.32 shows the packages and pin connections for two common logic devices, the 74LS00 (quad two-input NAND gate) and the 4001UBE (quad two-input NOR gate).

Table 10.8   Characteristics of common logic families

Characteristic	Logic family			
	74	74LS	74HC	40BE
Maximum supply voltage	5.25 V	5.25 V	5.5 V	18 V
Minimum supply voltage	4.75 V	4.75 V	4.5 V	3 V
Static power dissipation (mW per gate at 100 kHz)	10	2	negligible	negligible
Dynamic power dissipation (mW per gate at 100 kHz)	10	2	0.2	0.1
Typical propagation delay (ns)	10	10	10	105
Maximum clock frequency (MHz)	35	40	40	12
Speed–power product (pJ at 100 kHz)	100	20	1.2	11
Minimum output current (mA at $V_o = 0.4$ V)	16	8	4	1.6
Fan-out (number of LS loads that can be driven)	40	20	10	4
Maximum input current (mA at $V_i = 0.4$ V)	−1.6	−0.4	0.001	−0.001

4001UBE   **74LS00**

Figure 10.32   Packages and pin connections for two common logic devices

Circuit symbols introduced in this chapter

Figure 10.33

Figure 10.34 Symbols for R-S, D-type and J-K bistables

Formulae introduced in this chapter

Noise margin:
(page 172)

Noise margin $= V_{oh(MIN)} - V_{ih(MIN)}$
Noise margin $= V_{ol(MAX)} - V_{il(MAX)}$

Figure 10.35

Problems

10.1 Show how a four-input AND gate can be made from three two-input AND gates.

10.2 Show how a four-input OR gate can be made from three two-input OR gates.

10.3 Construct the truth table for the logic gate arrangement shown in Fig. 10.35.

10.4 Using only two-input NAND gates, show how each of the following logical functions can be satisfied:

  (a) two-input AND;
  (b) two-input OR;
  (c) four-input AND.

In each case, use the minimum number of gates.
(Hint: a two-input NAND gate can be made into an inverter by connecting its two inputs together)

A	B	C	Y
0	0	0	0
0	0	1	1
0	1	0	0
0	1	1	1
1	0	0	0
1	0	1	1
1	1	0	1
1	1	1	1

Figure 10.36

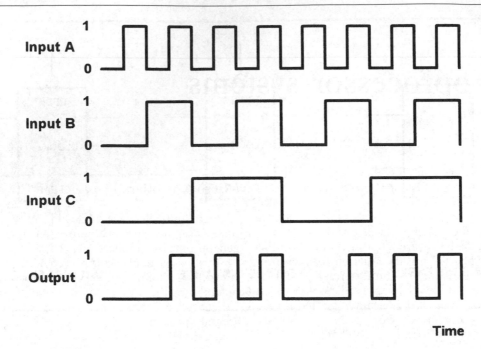

Figure 10.37

10.5   The rocket motor of a missile will operate if, and only if, the following conditions are satisfied:

(i) 'launch' signal is at logic 1;
(ii) 'unsafe height' signal is at logic 0;
(iii) 'target lock' signal is at logic 1.

Devise a suitable logic arrangement that will satisfy this requirement. Use the minimum number of logic gates.

10.6   An automatic sheet metal guillotine will operate if the following conditions are satisfied:

(i) 'guard lowered' signal is at logic 1;
(ii) 'feed jam' signal is at logic 0;
(iii) 'manual start' signal is at logic 1.

The sheet metal guillotine will also operate if the following conditions are satisfied:

(i) 'manual start' signal is at logic 1;
(ii) 'test key' signal is at logic 1.

Devise a suitable logic arrangement that will satisfy this requirement. Use the minimum number of logic gates.

10.7   Devise a logic arrangement using no more than four two-input gates that will satisfy the truth table shown in Fig. 10.36.

10.8   Devise a logic arrangement that will produce the output waveform from the three input waveforms shown in Fig. 10.37.

10.9   A logic device is marked '74LS90 2789'. To which family and sub-family of logic does it belong? When was the device manufactured?

10.10   A logic family recognizes a logic 1 input as being associated with any voltage between 2.0 V and 5.5 V. The same family produces an output in the range 2.6 V to 5.0 V corresponding to a logic 1 output. Determine the noise margin.

(Answers to these problems can be found on page 261.)

11

Microprocessor systems

Many of today's complex electronic systems are based on the use of a microprocessor or microcontroller. Such systems comprise hardware that is controlled by software. If it is necessary to change the way that the system behaves it is the software – rather than the hardware – that is changed. In this chapter we provide an introduction to microprocessors and explain, in simple terms, both how they operate and how they are used. We shall start by explaining some of the terminology that is used to describe different types of system that involve the use of a microprocessor or a similar device.

Microprocessor systems

Microprocessor systems are usually assembled on a single PCB comprising a microprocessor CPU together with a number of specialized support chips. These very large scale integrated (VLSI) devices provide input and output to the system, control and timing as well as storage for programs and data. Typical applications for microprocessor systems include the control of complex industrial processes. Typical examples are based on families of chips such as the Z80CPU plus Z80PIO, Z80CTC, and Z80SIO.

Single-chip microcomputers

A single-chip microcomputer is a complete computer system (comprising CPU, RAM and ROM, etc.) in a single VLSI package. A single-chip microcomputer requires very little external circuitry in order to provide all of the functions associated with a complete computer system (but usually with limited input and output capability). Single-chip microcomputers may be programmed using in-built programmable memories or via external memory chips. Typical applications of single-chip microcomputers include computer printers, instrument controllers, and displays. A typical example is the Z84C.

Microcontrollers

A microcontroller is a single-chip microcomputer that is designed specifically for control rather than general purpose applications. They are often used to satisfy a particular control requirement, such as controlling a motor drive. Single-chip microcomputers, on the other hand, usually perform a variety of different functions and may control several processes at the same time. Typical applications include control of peripheral devices such as motors, drives, printers, and minor subsystem components. Typical examples are the Z86E, 8051, 68705, and 89C51.

PIC microcontrollers

A PIC microcontroller is a general purpose microcontroller device that is normally used in a stand-alone application to perform simple logic, timing and input/output control. PIC devices provide a flexible low-cost solution that very effectively bridges the gap between single-chip computers and the use of discrete logic and timer chips. A number of PIC and microcontroller devices have been produced that incorporate a high-level language interpreter. The resident interpreter allows developers to develop their program languages such as BASIC rather than having to resort to more complex assembly language. This feature makes PIC microcontrollers very easy to use. PIC microcontrollers are used in 'self-contained' applications involving logic, timing, and simple analogue to digital and digital to analogue conversion. Typical examples are the PIC12C508 and PIC16C620.

Programmed logic devices

Whilst not an example of a microprocessor device, a programmed logic device (PLD) is a programmable chip that can carry out complex logical operations. For completeness, we have included a reference to such devices here. PLDs are capable of replacing a large number of conventional logic gates, thus minimizing chip count and reducing printed circuit board sizes. Programming is relatively straightforward and simply requires the derivation of complex logic functions using Boolean algebra (see page 285) or truth tables (see page 283). Typical examples are the 16L8 and 22V10.

Programmable logic controllers

Programmable logic controllers (PLCs) are microprocessor-based systems that are used for controlling a wide variety of automatic processes, from operating an airport baggage handling system to brewing a pint of your favourite lager. PLCs are rugged and modular and they are designed specifically for operation in the process control environment. The control program for a PLC is usually stored in one or more semiconductor memory devices. The program can be entered (or modified) by means of a simple hand-held programmer, a laptop controller, or downloaded over a local area network (LAN). PLC manufacturers include Allen Bradley, Siemens and Mitsubishi.

Microprocessor systems

The basic components of any microprocessor system are:

(a) a central processing unit (CPU);
(b) a memory, comprising both 'read/write' and 'read-only' devices (commonly called RAM and ROM respectively);

Figure 11.1   A microprocessor chip. The Z80A shown here is a popular 8-bit microprocessor that operates with a 4 MHz clock

(c) a means of providing input and output (I/O). For example, a keypad for input and a display for output.

In a microprocessor system the functions of the CPU are provided by a single very large scale integrated (VLSI) microprocessor chip (see Fig. 11.1). This chip is equivalent to many thousands of individual transistors. Semiconductor devices are also used to provide the read/write and read-only memory. Strictly speaking, both types of memory permit 'random access' since any item of data can be retrieved with equal ease regardless of its actual location within the memory. Despite this, the term 'RAM' has become synonymous with semiconductor **read/write memory**.

The basic components of the system (CPU, RAM, ROM and I/O) are linked together using a

Figure 11.2   Block diagram of a microprocessor system

multiple-wire connecting system known as a **bus** (see Fig. 11.2). Three different buses are present, these are:

(a) the **address bus** used to specify memory locations;
(b) the **data bus** on which data is transferred between devices; and
(c) the **control bus** which provides timing and control signals throughout the system.

The number of individual lines present within the address bus and data bus depends upon the particular microprocessor employed. Signals on all lines, no matter whether they are used for address, data, or control, can exist in only two basic states: logic 0 (**low**) or logic 1 (**high**). Data and addresses are represented by **binary numbers** (a sequence of 1s and 0s) that appear respectively on the data and address bus.

Many microprocessors designed for control and instrumentation applications make use of an 8-bit data bus and a 16-bit address bus. The largest binary number that can appear on an 8-bit data bus corresponds to the condition when all eight lines are at logic 1. Therefore the largest value of data that can be present on the bus at any instant of time is equivalent to the binary number 11111111 (or 255). Similarly, the highest address that can appear on a 16-bit address bus is 1111111111111111 (or 65 535). The full range of data values and addresses for a simple microprocessor of this type is thus:

Data   00000000 to 11111111
Addresses   0000000000000000 to 1111111111111111

Data representation

Binary numbers – particularly large ones – are not very convenient. To make numbers easier to handle we often convert binary numbers to **hexadecimal** (base 16). This format is easier for mere humans to comprehend and offers the advantage over denary (base 10) in that it can be converted to and from binary with ease. The first 16 numbers in binary, denary, and hexadecimal are shown in the table below. A single hexadecimal character (in the range zero to F) is used to represent a group of four binary digits (bits). This group of four bits (or single hex character) is sometimes called a **nibble**.

A **byte** of data comprises a group of 8 bits. Thus a byte can be represented by just two hexadecimal (**hex**) characters. A group of 16 bits (a **word**) can be represented by four hex characters, 32 bits (a **double word**) by eight hex characters, and so on.

The value of a byte expressed in binary can be easily converted to hex by arranging the bits in groups of four and converting each nibble into hexadecimal using the table shown below:

Binary (base 2)	Denary (base 10)	Hexadecimal (base 16)
0000	0	0
0001	1	1
0010	2	2
0011	3	3
0100	4	4
0101	5	5
0110	6	6
0111	7	7
1000	8	8
1001	9	9
1010	10	A
1011	11	B
1100	12	C
1101	13	D
1110	14	E
1111	15	F

Example 11.1

Convert hexadecimal A3 into binary.

Solution

From the table above, A = 1010 and 3 = 0011.
Thus A3 in hexadecimal is equivalent to 10100011 in binary.

Example 11.2

Convert binary 11101000 to hexadecimal.

Solution

From the table above, 1110 = E and 1000 = 8.
Thus 11101000 in binary is equivalent to E8 in hexadecimal.

Note that, to avoid confusion about whether a number is hexadecimal or decimal, we often place a $ symbol before a hexadecimal number or add an H to the end of the number. For example, 64 means decimal 'sixty-four'; whereas $64 means hexadecimal 'six-four', which is equivalent to decimal 100. Similarly, 7FH means hexadecimal 'seven-F', which is equivalent to decimal 127.

Data types

A byte of data can be stored at each address within the total memory space of a microprocessor system. Hence 1 byte can be stored at each of the 65 536 memory locations within a microprocessor system having a 16-bit address bus. Individual bits within a byte are numbered from 0 (least significant bit) to 7 (most significant bit). In the case of 16-bit words, the bits are numbered from 0 (least significant bit) to 15 (most significant bit). Negative (or **signed**) numbers can be represented using **two's complement** notation where the leading (most significant) bit indicates the sign of the number (1 = negative, 0 = positive).

The range of integer data values that can be represented as **bytes**, **words** and **long words** is shown in the table below:

Data type	Bits	Range of values
Unsigned byte	8	0 to 255
Signed byte	8	-128 to $+127$
Unsigned word	16	0 to 65 535
Signed word	16	$-32\,768$ to $+32\,767$

Data storage

The **semiconductor ROM** within a microprocessor system provides storage for the program code as well as any permanent data that requires storage. All of this data is referred to as **non-volatile** because it remains intact when the power supply is disconnected.

The **semiconductor RAM** within a microprocessor system provides storage for the transient data and variables that are used by programs. Part of the RAM is also used by the microprocessor as a temporary store for data whilst carrying out its normal processing tasks.

It is important to note that any program or data stored in RAM will be lost when the power supply is switched off or disconnected. The only exception to this is CMOS RAM that is kept alive by means of a small battery. This **battery-backed** memory is used to retain important data, such as the time and date.

When expressing the amount of storage provided by a memory device we usually use **Kilobytes** (Kbyte). It is important to note that a Kilobyte of memory is actually 1024 bytes (not 1000 bytes). The reason for choosing the Kbyte rather than the kbyte (1000 bytes) is that 1024 happens to be the nearest power of 2 (note that $2^{10} = 1024$).

The capacity of a semiconductor ROM is usually specified in terms of an **address range** and the number of bits stored at each address. For example, 2 K $\times$ 8 bits (capacity 2 Kbytes), 4 K $\times$ 8 bits (capacity 4 Kbytes), and so on. Note that it is not necessary for the entire memory space of a microprocessor to be populated by memory devices.

The microprocessor

The microprocessor **central processing unit** (CPU) forms the heart of any microprocessor or microcomputer system and, consequently, its operation is crucial to the entire system. The primary function of the microprocessor is that of fetching, decoding, and executing instructions resident in memory. As such, it must be able to transfer data from external memory into its own internal registers and vice versa. Furthermore, it must operate predictably, distinguishing, for example, between an operation contained within an instruction and any accompanying addresses of read/write memory locations. In addition, various system housekeeping tasks need to be performed including responding to interrupts from external devices.

The main parts of a microprocessor CPU are:

(a) **registers** for temporary storage of addresses and data;
(b) an **arithmetic logic unit** (ALU) that performs arithmetic and logic operations; and
(c) a means of controlling and timing operations within the system.

Figure 11.3 shows the principal internal features of a typical 8-bit microprocessor. We will briefly explain each of these features in turn.

Accumulator

The accumulator functions as a source and destination register for many of the basic microprocessor operations. As a **source register** it contains the data that will be used in a particular operation whilst as a **destination register** it will be used to hold the result of a particular operation. The accumulator (or **A-register**) features in a very large number of microprocessor operations, consequently more reference is made to this register than any others.

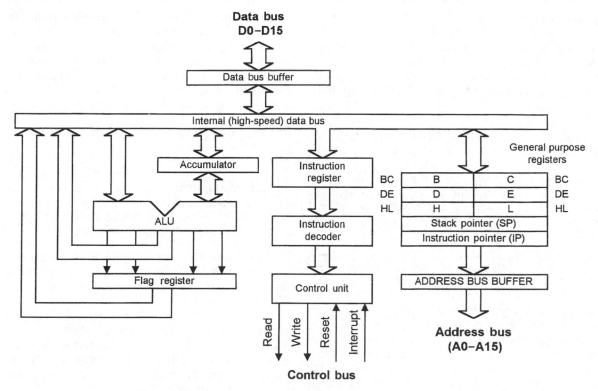

Figure 11.3   Internal arrangement of a typical 8-bit microprocessor

Instruction register

The instruction register provides a temporary storage location in which the current microprocessor instruction is held whilst it is being decoded. Program instructions are passed into the microprocessor, one at time, through the data bus. On the first part of each **machine cycle**, the instruction is fetched and decoded. The instruction is executed on the second (and subsequent) machine cycles. Each machine cycle takes a finite time (usually less than a microsecond) depending upon the frequency of the microprocessor's clock.

Data bus (D0 to D7)

The external data bus provides a highway for data that links all of the system components (such as random access memory, read-only memory, and input/output devices) together. In an 8-bit system, there are eight data lines, labelled D0 (the **least significant bit**) to D7 (the **most significant bit**) and data is moved around in groups of eight bits, or **bytes**.

Data bus buffer

The data bus buffer is a temporary register through which bytes of data pass on their way into, and out of, the microprocessor. The buffer is thus referred to as **bi-directional** with data passing out of the microprocessor on a **write operation** and into the processor during a **read operation**. The direction of data transfer is determined by the **control unit** as it responds to each individual program instruction.

Internal data bus

The internal data bus is a high-speed data highway that links all of the microprocessor's internal elements together. Data is constantly flowing backwards and forwards along the internal data bus lines.

General purpose registers

Many microprocessor operations (for example, adding two 8-bit numbers together) require the use of more than one register. There is also a requirement for temporarily storing the partial result of an operation whilst other operations take place. Both of these needs can be met by providing a number of general purpose registers. The use to which these registers are put is left mainly up to the programmer.

Stack pointer

When the time comes to suspend a particular task in order to briefly attend to something else, most microprocessors make use of a region of external random access memory (RAM) known as a **stack**. When the main program is interrupted, the microprocessor temporarily places in the stack the contents of its internal registers together with the address of the next instruction in the main program. When the interrupt has been attended to, the microprocessor recovers the data that has been stored temporarily in the stack together with the address of the next instruction within the main program. It is thus able to return to the main program exactly where it left off and with all the data preserved in its registers. The stack pointer is simply a register that contains the address of the last used stack location.

Program counter

Programs consist of a sequence of instructions that are executed by the microprocessor. These instructions are stored in external random access memory (RAM) or read-only memory (ROM). Instructions must be fetched and executed by the microprocessor in a strict sequence. By storing the address of the next instruction to be executed, the program counter allows the microprocessor to keep track of where it is within the program. The program counter is automatically incremented when each instruction is executed.

Address bus buffer

The address bus buffer is a temporary register through which addresses (in this case comprising 16 bits) pass on their way out of the microprocessor. In a simple microprocessor, the address buffer is unidirectional with addresses placed on the address bus during both read and write operations. The address bus lines are labelled A0

to A15, where A0 is the **least-significant** address bus line and A16 is the **most significant** address bus line. Note that a 16-bit address bus can be used to communicate with 65 536 individual memory locations. At each location a single **byte** of data is stored.

Control bus

The control bus is a collection of signal lines that are both used to control the transfer of data around the system and also to interact with external devices. The control signals used by microprocessors tend to differ with different types; however, the following are commonly found:

READ   an output signal from the microprocessor that indicates that the current operation is a **read operation**

WRITE   an output signal from the microprocessor that indicates that the current operation is a **write operation**

RESET   a signal that resets the internal registers and initializes the program counter so that the program can be restarted from the beginning

IRQ   interrupt request from an external device attempting to gain the attention of the microprocessor (the request may be honoured or ignored according to the state of the microprocessor at the time that the interrupt request is received)

NMI   non-maskable interrupt (i.e. an interrupt signal that cannot be ignored by the microprocessor)

Address bus (A0 to A15)

The address bus provides a highway for addresses that links with all of the system components (such as random access memory, read-only memory, and input/output devices). In a system with a 16-bit address bus, there are 16 address lines, labelled A0 (the **least significant bit**) to A15 (the **most significant bit**).

Instruction decoder

The instruction decoder is nothing more than an arrangement of logic gates that acts on the bits stored in the instruction register and determines which instruction is currently being referenced. The instruction decoder provides output signals for the microprocessor's control unit.

Control unit

The control unit is responsible for organizing the orderly flow of data within the microprocessor as well as generating, and responding to, signals on the control bus. The control unit is also responsible for the timing of all data transfers. This process is synchronized using an internal or external clock signal (not shown in Fig. 11.3).

Arithmetic logic unit (ALU)

As its name suggests, the ALU performs arithmetic and logic operations. The ALU has two inputs (in this case these are both 8 bits wide). One of these inputs is derived from the accumulator whilst the other is taken from the internal data bus via a temporary register (not shown in Fig. 11.3). The operations provided by the ALU usually include addition, subtraction, logical AND, logical

OR, logical exclusive-OR, shift left, shift right, etc. The result of most ALU operations appears in the accumulator.

Flag register (or status register)

The result of an ALU operation is sometimes important in determining what subsequent action takes place. The flag register contains a number of individual bits that are **set** or **reset** according to the outcome of an ALU operation. These bits are referred to as flags. The following flags are available in most microprocessors:

ZERO   the zero flag is set when the result of an ALU operation is zero (i.e. a byte value of 00000000)

CARRY   the carry flag is set whenever the result of an ALU operation (such as addition) generates a carry bit (in other words, when the result cannot be contained within an 8-bit register)

INTERRUPT   the interrupt flag indicates whether external interrupts are currently enabled or disabled

Figure 11.4   (a) A typical microprocessor clock circuit (b) waveform generated by the clock circuit shown in (a)

Clocks

The clock used in a microprocessor system is simply an accurate and stable square wave generator. In most cases the frequency of the square wave generator is determined by a quartz crystal. A simple 4 MHz square wave clock oscillator (together with the clock waveform that it produces) is shown in Fig. 11.4. Note that one complete clock cycle is sometimes referred to as a **T-state**.

Microprocessors sometimes have an internal clock circuit in which case the quartz crystal (or other resonant device) is connected directly to pins on the microprocessor chip. In Fig. 11.5(a) an external clock is shown connected to a microprocessor whilst in Fig. 11.5(b) an internal clock oscillator is used.

Microprocessor operation

The majority of operations performed by a microprocessor involve the movement of data.

Indeed, the program code (a set of instructions stored in ROM or RAM) must itself be fetched from memory prior to execution. The microprocessor thus performs a continuous sequence of instruction fetch and execute cycles. The act of fetching an instruction code (or operand or data value) from memory involves a **read operation** whilst the act of moving data from the microprocessor to a memory location involves a **write operation** – see Fig. 11.6.

Each cycle of CPU operation is known as a **machine cycle**. Program instructions may require several machine cycles (typically between two and five). The first machine cycle in any cycle consists of an **instruction fetch** (the instruction code is read from the memory) and it is known as the **M1 cycle**. Subsequent cycles, M2, M3, and so on, depend on the type of instruction that is being executed. This **fetch–execute sequence** is shown in Fig. 11.7.

Microprocessors determine the source of data (when it is being read) and the destination of data

(a)

(b)

Figure 11.5   (a) An external CPU clock arrangement
(b) an internal CPU clock arrangement

Figure 11.6   (a) A read operation (b) a write operation (when it is being written) by placing a unique address on the address bus. The address at which the data is to be placed (during a **write operation**) or from which it is to be fetched (during a **read operation**) can either constitute part of the memory of the system (in which case it may be within ROM or RAM) or it can be considered to be associated with input/output (I/O).

Since the data bus is connected to a number of VLSI devices, an essential requirement of such chips (e.g. ROM or RAM) is that their data outputs should be capable of being isolated from the bus whenever necessary. These chips are fitted with **select** or **enable** inputs that are driven by **address decoding logic** (not shown in Fig. 11.2). This logic ensures that ROM, RAM and I/O devices never simultaneously attempt to place data on the bus!

The inputs of the address decoding logic are derived from one, or more, of the address bus lines. The address decoder effectively divides the available memory into blocks corresponding to a particular function (ROM, RAM, I/O, etc.). Hence, where the processor is reading and writing to RAM, for example, the address decoding logic will ensure that only the RAM is selected whilst the ROM and I/O remain isolated from the data bus.

Within the CPU, data is stored in several registers. Registers themselves can be thought of as a simple pigeon-hole arrangement that can store as many bits as there are holes available. Generally, these devices can store groups of 16 or 32 bits. Additionally, some registers may be configured as either one register of 16 bits or two registers of 32 bits.

Some microprocessor registers are accessible to the programmer whereas others are used by the microprocessor itself. Registers may be classified as either general purpose or dedicated. In the latter case, a particular function is associated with the register, such as holding the result of an operation or signalling the result of a comparison. A typical microprocessor and its register model are shown in Fig. 11.8.

The arithmetic logic unit

The ALU can perform arithmetic operations (addition and subtraction) and logic (complementation, logical AND, logical OR, etc.). The ALU operates on two inputs (16 or 32 bits in length depending upon the CPU type) and it provides one output (again of 16 or 32 bits). In addition, the ALU status is preserved in the 'flag register' so that, for example, an overflow, zero or negative result can be detected. The control unit is responsible for the movement of data within the CPU and the management of control signals, both internal and external. The control unit asserts the requisite signals to read or write data as appropriate to the current instruction.

Input and output

The transfer of data within a microprocessor system involves moving groups of 8, 16 or 32 bits using the bus architecture described earlier. Consequently it is a relatively simple matter to transfer data into and out of the system in parallel form. This process is further simplified by using a programmable parallel I/O device (a Z80PIO, 8255, or equivalent VLSI chip). This device provides registers for the temporary storage of data that not only buffer the data but also provide a degree of electrical isolation from the system data bus.

Parallel data transfer is primarily suited to high-speed operation over relatively short distances, a typical example being the linking of a microcom-

Figure 11.7 A typical timing diagram for a fetch–execute cycle (note how the M1 cycle occupies four T-states)

Main register set

Accumulator (A)	Flags (F)
(B)	(C)
(D)	(E)
(H)	(L)

Special purpose registers

Interrupt vector (I)	Memory refresh (R)
Index register (IX)	
Index register (IY)	
Stack pointer (SP)	
Program counter (PC)	

Figure 11.8   The Z80 microprocessor showing some of its more important control signals and register set

puter to an adjacent dot matrix printer. There are, however, some applications in which parallel data transfer is inappropriate, the most common example being data communication by means of telephone lines. In such cases data must be sent serially (one bit after another) rather than in parallel form.

To transmit data in serial form, the parallel data from the microprocessor must be reorganized into a stream of bits. This task is greatly simplified by using an LSI interface device that contains a shift register that is loaded with parallel data from the data bus. This data is then read out as a serial bit stream by successive shifting. The reverse process, serial-to-parallel conversion, also uses a shift register. Here data is loaded in serial form, each bit shifting further into the register until it becomes full. Data is then placed simultaneously on the parallel output lines. The basic principles of

parallel-to-serial and serial-to-parallel data conversion are illustrated in Fig. 11.9.

An example program

The following example program is written in assembly code. The program transfers 8-bit data from an input port (Port A), complements (i.e. inverts) the data (by changing 0s to 1s and 1s to 0s in every bit position) and then outputs the result to an output port (Port B). The program repeats indefinitely.

Just three microprocessor instructions are required to carry out this task together with a fourth (jump) instruction that causes the three instructions to be repeated over and over again. A program of this sort is most easily written in **assembly code** which consists of a series of easy to

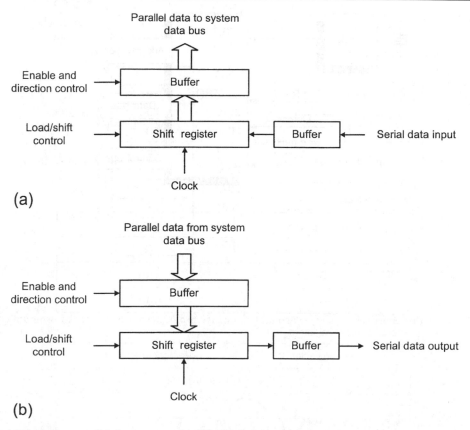

Figure 11.9 (a) Serial-to-parallel data conversion (b) parallel-to-serial data conversion

remember **mnemonics**. The **flowchart** for the program is shown in Fig. 11.10(a).

Address	Data	Assembly code	Comment
2000	DB FF	IN A, (FFH)	Get a byte from Port A
2002	2F	CPL	Invert the byte
2003	D3 FE	OUT (FEH), A	Output the byte to Port B
2005	C3 00 20	JP 2000	Go round again

The program occupies a total of 8 bytes of memory, starting at a hexadecimal address of 2000 as shown in Fig. 11.10(b). You should also note that the two ports, A and B, each have unique addresses; Port A is at hexadecimal address FF whilst Port B is at hexadecimal address FE.

Interrupts

A program that simply executes a loop indefinitely has a rather limited practical application. In most microprocessor systems we want to be able to interrupt the normal sequence of program flow in order to alert the microprocessor to the need to do something. We can do this with a signal known as an **interrupt**. There are two types of interrupt: **maskable** and **non-maskable**.

When a **non-maskable interrupt** input is asserted, the processor must suspend execution of the current instruction and respond immediately to the interrupt. In the case of a **maskable interrupt**, the processor's response will depend upon whether interrupts are currently enabled or disabled (when enabled, the CPU will suspend its current task and carry out the requisite interrupt service routine). The response to interrupts can be enabled or disabled by means of appropriate program instructions.

In practice, interrupt signals may be generated from a number of sources and since each will require its own customized response a mechanism must be provided for identifying the source of the interrupt and calling the appropriate interrupt service routine. In order to assist in this task, the

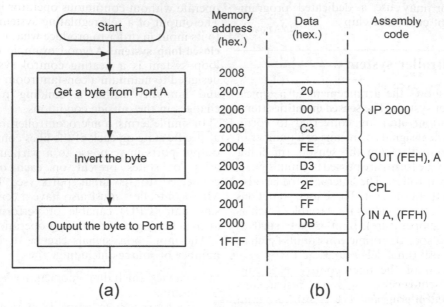

Figure 11.10 (a) A flowchart for the simple assembly language program (b) the 8 bytes of program code stored in memory

Figure 11.11 A microcontroller system

microprocessor may use a dedicated programmable interrupt controller chip.

A microcontroller system

Figure 11.11 shows the arrangement of a typical microcontroller system. The **sensed quantities** (temperature, position, etc.) are converted to corresponding electrical signals by means of a number of **sensors**. The outputs from the sensors (in either digital or analogue form) are passed as input signals to the microcontroller. The microcontroller also accepts inputs from the user. These **user set options** typically include target values for variables (such as desired room temperature), limit values (such as maximum shaft speed), or time constraints (such as 'on' time and 'off' time, delay time, etc.).

The operation of the microcontroller is controlled by a sequence of software instructions known as a **control program**. The control program operates continuously, examining inputs from sensors, user settings, and time data before making changes to the output signals sent to one or more controlled devices.

The **controlled quantities** are produced by the controlled devices in response to output signals from the microcontroller. The controlled device generally converts energy from one form into energy in another form. For example, the controlled device might be an electrical heater that converts electrical energy from the a.c. mains supply into heat energy thus producing a given temperature (the controlled quantity).

In most real-world systems there is a requirement for the system to be automatic or **self-regulating**. Once set, such systems will continue to operate without continuous operator intervention. The output of a self-regulating system is fed back to its input in order to produce what is known as a **closed-loop system**. A good example of a closed-loop system is a heating control system that is designed to maintain a constant room temperature and humidity within a building regardless of changes in the outside conditions.

In simple terms, a microcontroller must produce a specific state on each of the lines connected to its output ports in response to a particular combination of states present on each of the lines connected to its input ports (see Fig. 11.11). Microcontrollers must also have a central processing unit (CPU) capable of performing simple arithmetic, logical and timing operations.

The input port signals can be derived from a number of sources including:

(a) switches (including momentary action push-buttons);
(b) sensors (producing logic-level compatible outputs);
(c) keypads (both encoded and unencoded types).

The output port signals can be connected to a number of devices including:

(a) switches (including momentary action push-buttons);
(b) sensors (producing logic-level compatible outputs);
(c) keypads (both encoded and unencoded types).

Input devices

Input devices supply information to the computer system from the outside world. In an ordinary

Figure 11.12 An analogue input signal can be connected to a microcontroller input port via an analogue-to-digital converter (ADC)

Figure 11.13   An analogue output signal can be produced by connecting a digital-to-analogue converter (DAC) to a microcontroller output port

personal computer, the most obvious input device is the keyboard. Other input devices available on a PC are the mouse (pointing device), scanner and modem. Microcontrollers use much simpler input devices. These need be nothing more than individual switches or contacts that make and break but many other types of device are also used including many types of sensor that provide logic-level outputs (such as float switches, proximity detectors, light sensors, etc.).

It is important to note that, in order to be connected directly to the input port of a microcontroller, an input device must provide a logic compatible signal. This is because microcontroller inputs can only accept digital input signals with the same voltage levels as the logic power source. The 0 V ground level (often referred to as V_{SS} in the case of a CMOS microcontroller) and the positive supply (V_{DD} in the case of a CMOS microcontroller) is invariably 5 V ± 5%. A level of approximately 0 V indicates a logic 0 signal and a voltage approximately equal to the positive power supply indicates a logic 1 signal.

Other input devices may sense analogue quantities (such as velocity) but use a digital code to represent their value as an input to the microcontroller system. Some microcontrollers provide an internal **analogue-to-digital converter** (ADC) in order to greatly simplify the connection of analogue sensors as input devices.

Output devices

Output devices are used to communicate information or actions from a computer system to the outside world. In a personal computer system, the

most common output device is the CRT (cathode ray tube) display. Other output devices include printers and modems. As with input devices, microcontroller systems often use much simpler output devices. These may be nothing more than LEDs, piezoelectric sounders, relays and motors. In order to be connected directly to the output port of a microcontroller, an output device must, once again, be able to accept a logic compatible signal.

Where analogue quantities (rather than simple digital on/off operation) are required at the output a **digital-to-analogue converter** (DAC) will be needed.

Interface circuits

Where input and output signals are not **logic compatible** (i.e. when they are outside the range of signals that can be connected directly to the microcontroller) some additional interface circuitry may be required in order to shift the voltage levels or to provide additional current drive. For example, a common range of interface circuits (**solid-state relays**) is available that will allow a microcontroller to be easily interfaced to an a.c. mains-connected load. It then becomes possible for a small microcontroller (operating from only a 5 V d.c. supply) to control a central heating system operating from a 240 V a.c. mains.

Problems

11.1   Convert 3A hexadecimal to binary.

11.2   Convert 11000010 binary to hexadecimal.

11.3 Which of the following numbers is the largest?

(a) 21H
(b) $13
(c) 33_{10}
(d) 11100_2

11.4 How many unique addresses are available to a microprocessor CPU that has a 20-bit address bus?

11.5 What is the largest data value that can appear on a 10-bit data bus?

11.6 A signed byte has a binary value of 11111111. What is this equivalent to when expressed as a decimal number?

11.7 The following fragment of assembly language code is executed using a microprocessor:

IN A, (FEH)
CPL
OUT (FFH), A

(a) What are the addresses of the input and output ports?
(b) If a data value of 10101111 appears at the input port what value will appear at the output port after the code has been executed?

(Answers to these problems will be found on page 261.)

12

The 555 timer

The 555 timer is without doubt one of the most versatile integrated circuit chips ever produced. Not only is it a neat mixture of analogue and digital circuitry but its applications are virtually limitless in the world of digital pulse generation. The device also makes an excellent case study for newcomers to electronics because it combines a number of important concepts and techniques.

To begin to understand how timer circuits operate, it is worth spending a few moments studying the internal circuitry of the 555 timer, see Fig. 12.1. Essentially, the device comprises two operational amplifiers (used as differential comparators – see page 144) together with an R-S bistable element (see page 166). In addition, an inverting buffer (see page 164) is incorporated so that an appreciable current can be delivered to a load.

The main features of the device are as follows:

Feature	Function
A	A potential divider comprising R_1, R_2 and R_3 connected in series. Since all three resistors have the same values the input voltage (V_{CC}) will be divided into thirds, i.e. one-third of V_{CC} will appear at the junction of R_2 and R_3 whilst two-thirds of V_{CC} will appear at the junction of R_1 and R_2.
B	Two operational amplifiers connected as comparators. The operational amplifiers are used to examine the voltages at the **threshold** and **trigger** inputs and compare these with the fixed voltages from the potential divider (two-thirds and one-third of V_{CC} respectively).
C	An R-S bistable stage. This stage can be either **set** or **reset** depending upon the output from the comparator stage. An external reset input is also provided.
D	An open-collector transistor switch. This stage is used to discharge an external capacitor by effectively shorting it out whenever the base of the transistor is driven positive.
E	An inverting power amplifier. This stage is capable of **sourcing** and **sinking** enough current (well over 100 mA in the case of a standard 555 device) to drive a small relay or another low-resistance load connected to the output.

Unlike standard TTL logic devices, the 555 timer can both **source** and **sink** current. It's worth taking a little time to explain what we mean by these two terms:

(a) When **sourcing** current, the 555's output (pin-3) is in the **high** state and current will then flow *out of* the output pin into the load and down to 0 V, as shown in Fig. 12.2(a).

(b) When **sinking** current, the 555's output (pin-3) is in the **low** state in which case current will flow from the positive supply ($+V_{cc}$) through the load and *into* the output (pin-3), as shown in Fig. 12.2(b).

Returning to Fig. 12.1, the single transistor switch, TR1, is provided as a means of rapidly discharging an external timing capacitor. Because the series chain of resistors, R1, R2 and R3, all have identical values, the supply voltage (V_{CC}) is divided equally across the three resistors. Hence the voltage at the non-inverting input of IC1 is one-third of the supply voltage (V_{CC}) whilst that at the inverting input of IC2 is two-thirds of the supply voltage (V_{CC}). Hence if V_{CC} is 9 V, 3 V will appear at each resistor and the upper comparator will have 6 V applied to its inverting input whilst the lower comparator will have 3 V at its non-inverting input.

The 555 family

The standard 555 timer is housed in an 8-pin DIL package and operates from supply rail voltages of between 4.5 V and 15 V. This, of course, encompasses the normal range for TTL devices (5 V ± 5%) and thus the device is ideally suited for use with TTL circuitry.

The following versions of the standard 555 timer are commonly available.

Low power (CMOS) 555

This device is a CMOS version of the 5455 timer that is both pin and function compatible with its

Figure 12.1   Internal arrangement of a 555 timer

(a)   **(b)**

Figure 12.2   Loads connected to the output of a 555 timer: (a) current sourced by the timer when the output is high (b) current sourced by the timer when the output is low

standard counterpart. By virtue of its CMOS technology the device operates over a somewhat wider range of supply voltages (2 V to 18 V) and consumes minimal operating current (120 μA)

typical for an 18 V supply. Note that, by virtue of the low-power CMOS technology employed, the device does not have the same output current drive as that possessed by its standard counterparts. It can, however, supply up to two standard TTL loads.

Dual 555 timer (e.g. NE556A)

This is a dual version of the standard 555 timer housed in a 14-pin DIL package. The two devices may be used entirely independently and share the same electrical characteristics as the standard 555.

Low-power (CMOS) dual 555 (e.g. ICM75561PA)

This is a dual version of the low-power CMOS 555 timer contained in a 14-pin DIL package. The two devices may again be used entirely independently and share the same electrical characteristics as the low-power CMOS 555.

Pin connecting details for the above devices can be found in Appendix 4.

Monostable pulse generator

Figure 12.3 shows a standard 555 timer operating as a **monostable pulse generator**. The monostable timing period (i.e. the time for which the output is high) is initiated by a falling edge trigger pulse applied to the **trigger** input (pin-2), see Fig. 12.4.

When this falling edge trigger pulse is received and falls below one-third of the supply voltage, the output of IC2 goes **high** and the bistable will be

Figure 12.4   Waveforms for monostable operation

placed in the **set** state. The inverted Q output of the bistable then goes **low**, TR1 is placed in the **off** (non-conducting) state and the output voltage (pin-3) goes high.

The capacitor, C, then charges through the series resistor, R, until the voltage at the threshold input reaches two-thirds of the supply voltage (V_{CC}). At this point, the output of the upper comparator changes state and the bistable is **reset**. The inverted Q output then goes high, TR1 is driven into conduction and the final output goes **low**. The device then remains in the inactive state until another falling trigger pulse is received.

The output waveform produced by the circuit of Fig. 12.3 is shown in Fig. 12.4. The waveform has the following properties:

Time for which output is high: $t_{on} = 1.1CR$

Recommended trigger pulse width: $t_{tr} < \frac{t_{on}}{4}$

where t_{on} and t_{tr} are in seconds, C is in farads and R is in ohms.

The period of the 555 monostable output can be changed very easily by simply altering the values of the timing resistor, R, and/or timing capacitor, C. Doubling the value of R will double the timing period. Similarly, doubling the value of C will double the timing period. Note, however, that the performance of the timer may become unpredictable when the values of these components are outside the recommended range:

$C = 470\,\text{pF}$ to $470\,\mu\text{F}$

$R = 1\,\text{k}\Omega$ to $3.3\,\text{M}\Omega$

The required values of C and R for any desired monostable timing period can be determined from the formula shown earlier or by using the graph shown in Fig. 12.5.

Figure 12.3   555 monostable configuration

Figure 12.5 Graph for determining values of C, t_{on} and R for a 555 operating in monostable mode. The dotted line shows how a 10 ms pulse will be produced when $C = 100$ nF and $R = 91$ kΩ (see Example 12.1)

Example 12.1

Design a timer circuit that will produce a 10 ms pulse when a negative-going trigger pulse is applied to it.

Solution

Using the circuit shown in Fig. 12.4, the value of the monostable timing period can be calculated from the formula:

$$t_{on} = 1.1CR$$

We need to choose an appropriate value for C that is in the range stated earlier. Since we require a fairly modest time period we will choose a mid-range value for C. This should help to ensure that

the value of R is neither too small nor too large. A value of 100 nF should be appropriate and should also be easy to obtain. Making R the subject of the formula and substituting for $C = 100$ nF gives:

$$R = \frac{t_{on}}{1.1C} = \frac{10\,\text{ms}}{1.1 \times 100\,\text{nF}} = \frac{10 \times 10^{-3}}{110 \times 10^{-9}}$$

from which:

$$R = \frac{10}{110} \times 10^6 = 0.091 \times 10^6\,\Omega \text{ or } 9.1\,\text{k}\Omega$$

Alternatively, the graph shown in Fig. 12.5 can be used.

Example 12.2

Design a timer circuit that will produce a +5 V output for a period of 60 s when a 'start' button is operated. The time period is to be aborted when a 'stop' button is operated.

Solution

For the purposes of this question we shall assume that the 'start' and 'stop' buttons both have normally-open (NO) actions (see page 63).

The value of the monostable timing period can be calculated from the formula:

$$t_{on} = 1.1CR$$

We need to choose an appropriate value for C that is in the range stated earlier. Since we require a fairly long time period we will choose a relatively large value of C in order to avoid making the value of R too high. A value of 100 µF should be appropriate and should also be easy to obtain. Making R the subject of the formula and substituting for $C = 100$ µF gives:

$$R = \frac{t_{on}}{1.1C} = \frac{60\,\text{s}}{1.1 \times 100\,\text{µF}} = \frac{60}{110 \times 10^{-6}}$$

from which:

$$R = \frac{60}{110} \times 10^6 = 0.545 \times 10^6\,\Omega \text{ or } 545\,\text{k}\Omega$$

In practice 560 kΩ (the nearest preferred value – see page 18) would be adequate.

The 'start' button needs to be connected between pin-2 and ground whilst the 'stop' button needs to be connected between pin-4 and ground. Each of the inputs requires a 'pull-up' resistor to ensure that the input is taken high when the switch is not being operated. The precise value of the

Figure 12.6 60 s timer (see Example 12.2)

'pull-up' resistor is unimportant and a value of 10 kΩ will be perfectly adequate in this application. The complete circuit of the 60 s timer is shown in Fig. 12.6.

Astable pulse generator

Figure 12.7 shows how the standard 555 can be used as an **astable pulse generator**. In order to understand how this circuit operates, assume that

Figure 12.7 555 astable configuration

the output (pin-3) is initially high and that TR1 is in the non-conducting state. The capacitor, C, will begin to charge with current supplied by means of series resistors, $R1$ and $R2$.

When the voltage at the **threshold** input (pin-6) exceeds two-thirds of the supply voltage the output of the upper comparator, IC1, will change state and the bistable will become **reset** due to voltage transition that appears at R. This, in turn, will make the inverted O output go **high**, turning TR1 at the same time. Due to the inverting action of the buffer, IC4, the final output (pin-3) will go **low**.

The capacitor, C, will now discharge, with current flowing through $R2$ into the collector of TR1. At a certain point, the voltage appearing at the **trigger** input (pin-2) will have fallen back to one-third of the supply voltage at which point the lower comparator will change state and the voltage transition at S will return the bistable to its original **set** condition. The inverted Q output then goes low, TR1 switches **off** (no longer conducting), and the output (pin-3) goes **high**. Thereafter, the entire **charge/discharge cycle** is repeated indefinitely.

The output waveform produced by the circuit of Fig. 12.7 is shown in Fig. 12.8. The waveform has the following properties:

Time for which output is high: $t_{\text{on}} = 0.693C(R_1 + R_2)$

Time for which output is low: $t_{\text{off}} = 0.693CR_2$

Period of output waveform: $t = t_{\text{on}} + t_{\text{off}} = 0.693C(R_1 + 2R_2)$

Figure 12.8   Waveforms for astable operation

Pulse repetition frequency: $\text{p.r.f.} = \dfrac{1.44}{C(R_1 + 2R_2)}$

Mark to space ratio: $\dfrac{t_{\text{on}}}{t_{\text{off}}} = \dfrac{R_1 + R_2}{R_2}$

Duty cycle: $\dfrac{t_{\text{on}}}{t_{\text{on}} + t_{\text{off}}} = \dfrac{R_1 + R_2}{R_1 + 2R_2} \times 100\%$

where t is in seconds, C is in farads, R_1 and R_2 are in ohms.

When $R_1 = R_2$ the duty cycle of the astable output from the timer can be found by letting $R = R_1 = R_2$. Hence:

$$\frac{t_{\text{on}}}{t_{\text{off}}} = \frac{R_1 + R_2}{R_2} = \frac{R + R}{R} = \frac{2}{1} = 2$$

In this case the duty cycle will be given by:

$$\frac{t_{\text{on}}}{t_{\text{on}} + t_{\text{off}}} = \frac{R_1 + R_2}{R_1 + 2R_2} \times 100\% = \frac{R + R}{R + 2R} \times 100\%$$

$$= \frac{2}{3} \times 100\% = 67\%$$

The p.r.f. of the 555 astable output can be changed very easily by simply altering the values of R_1, R_2, and C. The values chosen for R_1, R_2 and C should normally be selected from within the following ranges in order to provide satisfactory performance:

$C = 10\,\text{nF}$ to $470\,\mu\text{F}$

$R_1 = 1\,\text{k}\Omega$ to $1\,\text{M}\Omega$

$R_2 = 1\,\text{k}\Omega$ to $1\,\text{M}\Omega$

The required values of C, R_1 and R_2 for any required p.r.f. and duty cycle can be determined from the formulae shown earlier. Alternatively, the graph shown in Fig. 12.9 can be used when R_1 and R_2 are equal in value (corresponding to a 67% duty cycle).

Figure 12.9   Graph for determining values of C, p.r.f. and R_2 for a 555 operating in astable mode where $R_2 >> R_1$ (i.e. for square wave operation). The dotted line shows how a 50 Hz square wave will be produced when $C = 100\,\text{nF}$ and $R = 144\,\text{k}\Omega$ (see Example 12.4)

Square wave generators

Because the high time (t_{on}) is always greater than the low time (t_{off}), the mark to space ratio produced by a 555 timer can never be made equal to (or less than) unity. This could be a problem if we need to produce a precise square wave in which $t_{on} = t_{off}$. However, by making R_2 very much larger than R_1, the timer can be made to produce a reasonably symmetrical square wave output (note that the minimum recommended value for R_2 is 1 kΩ – see earlier).

If $R_2 \gg R_1$ the expressions for p.r.f. and the duty cycle simplify to:

$$\text{p.r.f.} \simeq \frac{0.72}{CR_2}$$

$$\frac{t_{on}}{t_{on} + t_{off}} \simeq \frac{R_2}{2R_2} \times 100\% = \frac{1}{2} \times 100\% = 50\%$$

Example 12.3

Design a pulse generator that will produce a p.r.f. of 10 Hz with a 67% duty cycle.

Solution

Using the circuit shown in Fig. 12.7, the value of p.r.f. can be calculated from:

$$\text{p.r.f.} = \frac{1.44}{C(R_1 + 2R_2)}$$

Since the specified duty cycle is 67% we can make R_1 equal to R_2. Hence if $R = R_1 = R_2$ we obtain the following relationship:

$$\text{p.r.f.} = \frac{1.44}{C(R + 2R)} = \frac{1.44}{3CR} = \frac{0.48}{CR}$$

We need to choose an appropriate value for C that is in the range stated earlier. Since we require a fairly low value of p.r.f. we will choose a value for C of 1 μF. This should help to ensure that the value of R is neither too small nor too large. A value of 1 μF should also be easy to obtain. Making R the subject of the formula and substituting for $C = 1$ μF gives:

$$R = \frac{0.48}{\text{p.r.f.} \times C} = \frac{0.48}{\text{p.r.f.} \times 1 \times 10^{-6}} = \frac{480 \times 10^3}{\text{p.r.f.}}$$

$$= \frac{480 \times 10^3}{100} = 4.8 \times 10^3 = 4.8 \text{ k}\Omega$$

Example 12.4

Design a 5 V 50 Hz square wave generator using a 555 timer.

Solution

Using the circuit shown in Fig. 12.7, when $R_2 \gg R_1$, the value of p.r.f. can be calculated from:

$$\text{p.r.f.} \simeq \frac{0.72}{CR_2}$$

We shall use the minimum recommended value for R_1 (i.e. 10 kΩ) and ensure that the value of R_2 that we calculate from the formula is at least ten times larger in order to satisfy the criteria that R_2 should be very much larger than R_1. When selecting the value for C we need to choose a value that will keep the value of R_2 relatively large. A value of 100 nF should be about right and should also be easy to locate. Making R_2 the subject of the formula and substituting for $C = 100$ nF gives:

$$R_2 = \frac{0.72}{\text{p.r.f.} \times C} = \frac{0.72}{50 \times 100 \times 10^{-9}} = \frac{0.72}{5 \times 10^{-6}}$$

$$= \frac{0.72 \times 10^6}{5} = 0.144 \times 10^6 = 144 \text{ k}\Omega$$

Alternatively, the graph shown in Fig. 12.9 can be used.

The value of R_2 is thus more than 100 times larger than the value that we are using for R_1. As a consequence the timer should produce a good square wave output.

The complete circuit of the 5 V 50 Hz square wave generator is shown in Fig. 12.10.

Figure 12.10   5 V 50 Hz square wave generator (see Example 12.4)

Figure 12.11 Variable pulse generator using two 555 timers

A variable pulse generator

Figure 12.11 shows how a variable pulse generator can be constructed using two 555 timers (or one 556 dual timer). The first timer, IC1, operates in astable mode whilst the second timer, IC2, operates in monostable mode. The p.r.f. is adjustable over the range 10 Hz to 10 kHz whilst pulse widths can be varied from 50 μs to 50 ms. The output is adjustable from 0 V to 10 V.

Important formulae introduced in this chapter

Monostable 555 timer
(page 194)

$$t_{on} = 1.1CR$$

Astable 555 timer
(page 196)

$$t_{on} = 0.693C(R_1 + R_2)$$

$$t_{off} = 0.693CR_2$$

$$t = 0.693C(R_1 + 2R_2)$$

$$\text{p.r.f.} = \frac{1.44}{C(R_1 + 2R_2)}$$

$$\frac{t_{on}}{t_{off}} = \frac{R_1 + R_2}{R_2}$$

$$\frac{t_{on}}{t_{on} + t_{off}} = \frac{R_1 + R_2}{R_1 + 2R_2} \times 100\%$$

When $R_2 \gg R_1$
(page 198)

$$\text{p.r.f.} \simeq \frac{0.72}{CR_2}$$

$$\frac{t_{on}}{t_{on} + t_{off}} \simeq 50\%$$

Problems

12.1 Design a timer circuit that will produce a 10 V 2 ms pulse when a 10 V negative-going trigger pulse is applied to it.

12.2 Design a timer circuit that will produce time periods that can be varied from 1 s to 10 s. The timer circuit is to produce a +5 V output.

12.3 Design a timer circuit that will produce a 67% duty cycle output at 400 Hz.

12.4   Design a timer circuit that will produce a square wave output at 1 kHz.

12.5   Refer to the variable pulse generator circuit shown in Fig. 12.10. Identify the component(s) that:
   (a) provides variable adjustment of pulse width;
   (b) provides decade range selection of pulse width;
   (c) limits the range of variable adjustment of pulse width;
   (d) provides variable adjustment of p.r.f.;
   (e) provides decade range selection of p.r.f.;
   (f) limits the range of variable adjustment of p.r.f.;
   (g) provides variable adjustment of output amplitude;
   (h) protects IC_2 against a short-circuit connected at the output;
   (i) removes any unwanted signals appearing on the supply rail;
   (j) forms the trigger pulse required by the monostable stage.

12.6   A 555 timer is rated for a maximum output current of 120 mA. What is the minimum value of load resistance that can be used if the device is to be operated from a 6 V d.c. supply?

(Answers to these problems will be found on page 261.)

13

Radio

Maxwell first suggested the existence of electromagnetic waves in 1864. Later, Heinrich Rudolf Hertz used an arrangement of rudimentary resonators to demonstrate the existence of electromagnetic waves. Hertz's apparatus was extremely simple and comprised two resonant loops, one for transmitting and the other for receiving. Each loop acted both as a tuned circuit and as a resonant aerial. The transmitting loop was excited by means of an induction coil and battery. Some of the energy radiated by the transmitting loop was intercepted by the receiving loop and the received energy was conveyed to a spark gap where it could be released as an arc. The energy radiated by the transmitting loop was in the form of an **electromagnetic wave** – a wave that has both electric and magnetic field components and that travels at the speed of light.

In 1894, Marconi demonstrated the commercial potential of the phenomenon that Maxwell predicted and Hertz actually used in his apparatus. It was also Marconi who made radio a reality by pioneering the development of telegraphy without wires (i.e. 'wireless'). Marconi was able to demonstrate very effectively that information could be exchanged between distant locations without the need for a 'land-line'.

Marconi's system of **wireless telegraphy** proved to be invaluable for maritime communications (ship to ship and ship to shore) and was to be instrumental in saving many lives. The military applications of radio were first exploited during the First World War (1914 to 1918) and, during that period, radio was first used in aircraft.

The radio frequency spectrum

Radio frequency signals are generally understood to occupy a frequency range that extends from a few tens of kilohertz (kHz) to several hundred gigahertz (GHz). The lowest part of the radio frequency range that is of practical use (below 30 kHz) is only suitable for narrow-band communication. At this frequency, signals propagate as ground waves (following the curvature of the earth) over very long distances. At the other extreme, the highest frequency range that is of practical importance extends above 30 GHz. At these microwave frequencies, considerable bandwidths are available (sufficient to transmit many television channels using point-to-point links or to permit very high definition radar systems) and signals tend to propagate strictly along line-of-sight paths.

At other frequencies signals may propagate by various means including reflection from ionized layers in the ionosphere. At frequencies between 3 MHz and 30 MHz ionospheric propagation regularly permits intercontinental broadcasting and communications.

For convenience, the radio frequency spectrum is divided into a number of bands, each spanning a decade of frequency. The use to which each frequency range is put depends upon a number of factors, paramount amongst which are the propagation characteristics within the band concerned. Other factors that need to be taken into account include the efficiency of practical aerial systems in the range concerned and the bandwidth available. It is also worth noting that, although it may appear from Fig. 13.1 that a great deal of the radio frequency spectrum is not used, it should be stressed that competition for frequency space is fierce. Frequency allocations are, therefore, ratified by international agreement and the various user services carefully safeguard their own areas of the spectrum.

Electromagnetic waves

As with light, radio waves propagate outwards from a source of energy (transmitter) and comprise electric (E) and magnetic (H) fields at right angles to one another. These two components, the **E-field** and the **H-field**, are inseparable. The resulting wave travels away from the source with the E and H lines mutually at right angles to the direction of **propagation**, as shown in Fig. 13.2.

Figure 13.1 The radio frequency spectrum

Radio waves are said to be **polarized** in the plane of the electric (E) field. Thus, if the E-field is vertical, the signal is said to be vertically polarized whereas if the E-field is horizontal, the signal is said to be horizontally polarized.

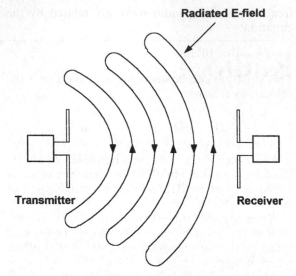

Figure 13.3 E-field lines between a transmitter and receiver

Figure 13.3 shows the electric E-field lines in the space between a transmitter and a receiver. The transmitter aerial (a simple dipole – see page 210) is supplied with a high frequency alternating current. This gives rise to an alternating electric field between the ends of the aerial and an alternating magnetic field around (and at right angles to) it.

The direction of the E-field lines is reversed on each cycle of the signal as the **wavefront** moves outwards from the source. The receiving aerial intercepts the moving field and voltage and current is induced in it as a consequence. This voltage and current is similar (but of smaller amplitude) to that produced by the transmitter.

Frequency and wavelength

Radio waves propagate in air (or space) at the **speed of light** (300 million metres per second). The velocity of propagation, v, wavelength, λ, and

Figure 13.2 Amplitude and frequency modulation

frequency, f, of a radio wave are related by the equation:

$$v = f\lambda = 3 \times 10^8 \text{ m/s}$$

This equation can be arranged to make f or λ the subject, as follows:

$$f = \frac{3 \times 10^8}{\lambda} \text{ Hz} \quad \text{and} \quad \lambda = \frac{3 \times 10^8}{f} \text{ m}$$

As an example, a signal at a frequency of 1 MHz will have a wavelength of 300 m whereas a signal at a frequency of 10 MHz will have a wavelength of 30 m.

When a radio wave travels in a cable (rather than in air or 'free space') it usually travels at a speed that is between 60% and 80% of that of the speed of light.

Example 13.1

Determine the frequency of a radio signal that has a wavelength of 15 m.

Solution

Using the formula $f = \dfrac{3 \times 10^8}{\lambda}$ Hz where $\lambda = 15$ m gives:

$$f = \frac{3 \times 10^8}{15} = \frac{300 \times 10^6}{15} = 20 \times 10^6 \text{ Hz or 20 MHz}$$

Example 13.2

Determine the wavelength of a radio signal that has a frequency of 150 MHz.

Solution

Using the formula $\lambda = \dfrac{3 \times 10^8}{f}$ m where $f = 150$ MHz gives:

$$\lambda = \frac{3 \times 10^8}{f} = \frac{3 \times 10^8}{150 \text{ MHz}} = \frac{3 \times 10^8}{150 \times 10^6} = \frac{300 \times 10^6}{150 \times 10^6}$$
$$= \frac{300}{150} = 2 \text{ m}$$

Example 13.3

If the wavelength of a 30 MHz signal in a cable is 8 m, determine the velocity of propagation of the wave in the cable.

Solution

Using the formula $v = f\lambda$, where v is the velocity of propagation in the cable, gives:

$$v = f\lambda = 30 \text{ MHz} \times 8 \text{ m}$$
$$= 240 \times 10^6 \text{ m/s} = 2.4 \times 10^8 \text{ m/s}$$

A simple CW transmitter and receiver

Figure 13.4 shows a simple radio communication system comprising a **transmitter** and **receiver** for use with **continuous wave** (CW) signals. Communication is achieved by simply switching (or 'keying') the radio frequency signal on and off. Keying can be achieved by interrupting the supply to the power amplifier stage or even the oscillator stage; however, it is normally applied within the driver stage that operates at a more modest power

Figure 13.4 Simplified block schematic of a complete radio communication system comprising AM transmitter and a TRF receiver

level. Keying the oscillator stage usually results in impaired frequency stability. On the other hand, attempting to interrupt the appreciable currents and/or voltages that appear in the power amplifier stage can also prove to be somewhat problematic.

The simplest form of CW receiver need consist of nothing more than a radio frequency amplifier (which provides gain and selectivity) followed by a detector and an audio amplifier. The **detector** stage mixes a locally generated radio frequency signal produced by the **beat frequency oscillator** (BFO) with the incoming signal to produce a signal within the audio frequency range.

As an example, assume that the incoming signal is at a frequency of 100 kHz and that the BFO is producing a signal at 99 kHz. A signal at the difference between these two frequencies (1 kHz) will appear at the output of the detector stage. This will then be amplified within the audio stage before being fed to the loudspeaker.

Example 13.4

A radio wave has a frequency of 162.5 kHz. If a beat frequency of 1.25 kHz is to be obtained, determine the two possible BFO frequencies.

Solution

The BFO can be above or below the incoming signal frequency by an amount that is equal to the beat frequency (i.e. the audible signal that results from the 'beating' of the two frequencies and which appears at the output of the detector stage).

Hence, $f_{BFO} = f_{RF} \pm f_{AF}$

from which:

$$f_{BFO} = 162.5 \text{ kHz} \pm 1.25 \text{ kHz}$$
$$= 160.75 \text{ kHz or } 163.25 \text{ kHz}$$

Morse code

Transmitters and receivers for CW operation are extremely simple but nevertheless they can be extremely efficient. This makes them particularly useful for disaster and emergency communications or for any situation that requires optimum use of low power equipment. Signals are transmitted using the code invented by Samuel Morse (see Fig. 13.5).

Morse code uses a combination of dots (short periods of transmission) and dashes (slightly

Figure 13.5   Morse code

Figure 13.6   RF signal for the Morse letter 'C'

longer periods of transmission) to represent characters. As an example, Fig. 13.6 shows how the radio frequency carrier is repeatedly switched on and off to transmit the character 'C'.

Modulation

In order to convey information using a radio frequency carrier, the signal information must be superimposed or 'modulated' onto the carrier. **Modulation** is the name given to the process of changing a particular property of the carrier wave in sympathy with the instantaneous voltage (or current) signal.

The most commonly used methods of modulation are **amplitude modulation** (AM) and **frequency** modulation (FM). In the former case, the carrier amplitude (its peak voltage) varies according to the voltage, at any instant, of the modulating signal. In the latter case, the carrier frequency is varied in accordance with the voltage, at any instant, of the modulating signal.

Figure 13.7 shows the effect of amplitude and frequency modulating a sinusoidal carrier (note that the modulating signal is, in this case, also

Figure 13.7   Amplitude modulation (AM) and frequency modulation (FM)

sinusoidal). In practice, many more cycles of the RF carrier would occur in the time span of one cycle of the modulating signal.

Demodulation

Demodulation is the reverse of modulation and is the means by which the signal information is recovered from the modulated carrier. Demodulation is achieved by means of a **demodulator** (sometimes called a **detector**). The output of a demodulator consists of a reconstructed version of the original signal information present at the input of the modulator stage within the transmitter. We shall see how this works a little later.

An AM transmitter

Figure 13.8 shows the block schematic of a simple AM transmitter. An accurate and stable RF oscillator generates the radio frequency **carrier** signal. The output of this stage is then amplified and passed to a modulated RF power amplifier stage. The inclusion of an amplifier between the RF oscillator and the modulated stage also helps to improve frequency stability.

The low-level signal from the microphone is amplified using an AF amplifier before it is passed to an AF power amplifier. The output of the

Figure 13.8   An AM transmitter

power amplifier is then fed as the supply to the modulated RF power amplifier stage. Increasing and reducing the supply to this stage is instrumental in increasing and reducing the amplitude of its RF output signal.

The modulated RF signal is then passed through an aerial tuning unit that matches the aerial to the RF power amplifier and also helps to reduce the level of any unwanted harmonic components that may be present.

An FM transmitter

Figure 13.9 shows the block schematic of a simple FM transmitter. Once again, an accurate and stable RF oscillator generates the radio frequency carrier signal. As with the AM transmitter, the output of this stage is amplified and passed to an RF power amplifier stage. Here again, the inclusion of an amplifier between the RF oscillator and the RF power stage helps to improve frequency stability.

The low-level signal from the microphone is amplified using an AF amplifier before it is passed to a **variable reactance** element (e.g. a variable capacitance diode – see page 86) within the RF

oscillator tuned circuit. The application of the AF signal to the variable reactance element causes the frequency of the RF oscillator to increase and decrease in sympathy with the AF signal.

The final RF signal from the power amplifier is passed through an aerial tuning unit that matches the aerial to the RF power amplifier and also helps to reduce the level of any unwanted harmonic components that may be present.

A tuned radio frequency (TRF) receiver

Tuned radio frequency (TRF) receivers provide a means of receiving local signals using fairly minimal circuitry. The simplified block schematic of a TRF receiver is shown in Fig. 13.10.

The signal from the aerial is applied to an RF amplifier stage. This stage provides a moderate amount of gain at the signal frequency. It also provides **selectivity** by incorporating one or more tuned circuits at the signal frequency. This helps the receiver to reject signals that may be present on adjacent channels.

The output of the RF amplifier stage is applied to the demodulator. This stage recovers the audio

Figure 13.9 An FM transmitter

Figure 13.10 A TRF receiver

frequency signal from the modulated RF signal. The demodulator stage may also incorporate a tuned circuit to further improve the selectivity of the receiver.

The output of the demodulator stage is fed to the input of the AF amplifier stage. This stage increases the level of the audio signal from the demodulator so that it is sufficient to drive a loudspeaker.

TRF receivers have a number of limitations with regard to sensitivity and selectivity and this makes them generally unsuitable for use in commercial radio equipment.

A superhet receiver

Superhet receivers provide both improved **sensitivity** (the ability to receive weak signals) and improved **selectivity** (the ability to discriminate signals on adjacent channels) when compared with TRF receivers. Superhet receivers are based on the **supersonic-heterodyne** principle where the wanted input signal is converted to a fixed **intermediate frequency** (IF) at which the majority of the gain and selectivity is applied. The intermediate frequency chosen is generally 455 kHz or 470 kHz for AM receivers and 10.7 MHz for communications and FM receivers. The simplified block schematic of a simple superhet receiver is shown in Fig. 13.11.

The signal from the aerial is applied to an **RF amplifier** stage. As with the TRF receiver, this stage provides a moderate amount of gain at the signal frequency. The stage also provides selectivity by incorporating one or more tuned circuits at the signal frequency.

The output of the RF amplifier stage is applied to the **mixer** stage. This stage combines the RF signal with the signal derived from the **local oscillator** stage in order to produce a signal at the **intermediate frequency** (IF). It is worth noting that the output signal produced by the mixer actually contains a number of signal components, including the sum and difference of the signal and local oscillator frequencies as well as the original signals plus harmonic components. The wanted signal (i.e. that which corresponds to the IF) is passed (usually by some form of filter) to the IF amplifier stage. This stage provides amplification as well as a high degree of selectivity.

The output of the IF amplifier stage is fed to the demodulator stage. As with the TRF receiver, this stage is used to recover the audio frequency signal from the modulated RF signal.

Finally, the AF signal from the demodulator stage is fed to the AF amplifier. As before, this stage increases the level of the audio signal from the demodulator so that it is sufficient to drive a loudspeaker.

In order to cope with a wide variation in signal amplitude, superhet receivers invariably incorporate some form of **automatic gain control** (AGC). In most circuits the d.c. level from the AM demodulator (see page 209) is used to control the gain of the IF and RF amplifier stages. As the signal level increases, the d.c. level from the demodulator stage increases and this is used to reduce the gain of both the RF and IF amplifiers.

The superhet receiver's intermediate frequency, f_{IF}, is the difference between the signal frequency,

Figure 13.11 A superhet receiver

f_{RF}, and the local oscillator frequency, f_{LO}. The desired local oscillator frequency can be calculated from the relationship:

$$f_{LO} = f_{RF} \pm f_{IF}$$

Note that in most cases (and in order to simplify tuning arrangements) the local oscillator operates above the signal frequency, i.e. $f_{LO} = f_{RF} + f_{IF}$.

Example 13.5

A VHF Band II FM receiver with a 10.7 MHz IF covers the signal frequency range 88 MHz to 108 MHz. Over what frequency range should the local oscillator be tuned?

Solution

Using $f_{LO} = f_{RF} + f_{IF}$ when $f_{RF} = 88$ MHz gives $f_{LO} = 88$ MHz $+ 10.7$ MHz $= 98.7$ MHz.

Using $f_{LO} = f_{RF} + f_{IF}$ when $f_{RF} = 108$ MHz gives $f_{LO} = 108$ MHz $+ 10.7$ MHz $= 118.7$ MHz.

The local oscillator tuning range should therefore be from 98.7 MHz to 118.7 MHz.

RF amplifiers

Figure 13.12 shows the circuit of a typical RF amplifier stage (this circuit can also be used as an IF amplifier in a superhet receiver). You might like to contrast this circuit with Fig. 7.33 shown on page 129. The amplifier operates in Class A and uses a small-signal NPN transistor connected in common-emitter mode. The essential difference between the circuit shown in Fig. 13.12 and that shown in Fig. 7.33 is that the RF amplifier uses a parallel tuned circuit as a collector load. To improve matching and prevent damping of the tuned circuit (which results in a reduction in Q-factor and selectivity) the collector of TR_1 is **tapped** into the tuned circuit rather than connected straight across it). Since the tuned circuit has maximum impedance at resonance (see page 74), maximum gain will occur at the resonant frequency. By using a tuned circuit with high Q-factor it is possible to limit the response of the amplifier to a fairly narrow range of frequencies. The output (to the next stage) is taken from a secondary winding, L_2, on the main tuning inductor, L_1.

In order to further improve **selectivity** (i.e. the ability to discriminate between signals on adjacent channels) several tuned circuits can be used together in order to form a more effective **band-pass filter**. Figure 13.13 shows one possible arrangement. When constructing an RF filter using several tuned circuits it is necessary to use the optimum coupling between the two tuned circuits. Figure 13.14 illustrates this point.

If the two tuned circuits are too 'loosely' coupled (they are said to be **under-coupled**) the frequency response characteristic becomes flat and insufficient output is obtained. On the other hand, if they are too 'tightly' coupled (they are said to be **over-coupled**) the response becomes broad and 'double-humped'. The optimum value of coupling (when the two tuned circuits are said to be **critically coupled**) corresponds to a frequency response that has a relatively flat top and steep sides.

Figure 13.12 An RF amplifier

Figure 13.13 A tuned bandpass filter

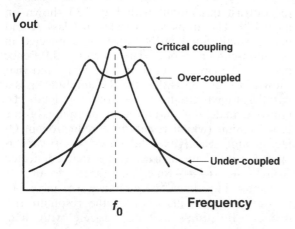

Figure 13.14 Frequency response for coupled tuned circuits showing different amounts of coupling between the tuned circuits

AM demodulators

Figure 13.15 shows the circuit of a typical AM demodulator stage. The RF input is applied to a parallel tuned circuit (L_1 and C_1) which exhibits a very high impedance at the signal frequency. A secondary coupling winding, L_2, is used to match the relatively low impedance of the **diode demodulator** circuit to the high impedance of the input tuned circuit. Diode D_1 acts as a half-wave rectifier conducting only on positive-going half-cycles of the radio frequency signal. Capacitor C_1 charges to the peak value of each positive-going half-cycle that appears at the cathode of D_1.

The voltage that appears across C_1 roughly follows the peak of the half-cycles of rectified voltage. R_1 and C_2 form a simple filter circuit to remove unwanted RF signal components (this circuit works in just the same way as the smoothing filter that we met in Chapter 6 – see pages 107 and 108). The final result is a voltage

waveform appearing across C_2 that resembles the original modulating signal. As well as providing a current path for D_1, R_2 forms a discharge path for C_1 and C_2. Coupling capacitor C_3 is used to remove any d.c. component from the signal that appears at the output of the demodulator. Waveforms for the demodulator circuit are shown in Fig. 13.16. Figure 13.17 shows a complete IF amplifier together with an AM demodulator stage. Circuits of this type are used in simple superhet receivers.

Aerials

We shall start by describing one of the most fundamental types of aerial, the **half-wave dipole**. The basic half-wave dipole aerial (Fig. 13.18)

Figure 13.16 Waveforms for the diode AM demodulator

Figure 13.15 A diode AM demodulator

Figure 13.17   A complete RF/IF amplifier and AM demodulator

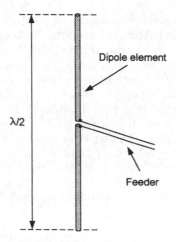

Figure 13.18   A half-wave dipole aerial

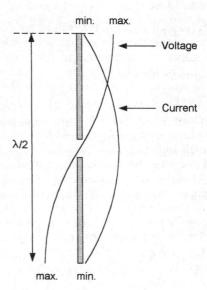

Figure 13.19   Voltage and current distribution for a half-wave dipole

consists of a single conductor having a length equal to one-half of the length of the wave being transmitted or received. The conductor is then split in the centre to enable connection to the feeder. In practice, because of the capacitance effects between the ends of the aerial and ground, the aerial is invariably cut a little shorter than a half wavelength.

The length of the aerial (from end to end) is equal to one half wavelength, hence $l = \dfrac{\lambda}{2}$.

Now since $v = f\lambda$ we can conclude that, for a half-wave dipole, $l = \dfrac{v}{2f}$.

Note that l is the **electrical length** of the aerial rather than its actual **physical length**. End effects or capacitance effects at the ends of the aerial require that we reduce the actual length of the aerial and a 5% reduction in length is typically required for an aerial to be resonant at the centre of its designed tuning range.

Figure 13.19 shows the distribution of current and voltage along the length of a half-wave dipole aerial. The current is maximum at the centre and zero at the ends. The voltage is zero at the centre and maximum at the ends. This implies that the impedance is not constant along the length of aerial but varies from a maximum at the ends (maximum voltage, minimum current) to a minimum at the centre.

The dipole aerial has directional properties illustrated in Fig. 13.20. Figure 13.20(a) shows the radiation pattern of the aerial in the plane of the antenna's electric field whilst Fig. 13.20(b)

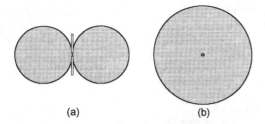

(a)   (b)

Figure 13.20 Radiation pattern for a half-wave dipole: (a) radiation in the electric plane (b) radiation in the magnetic plane

shows the radiation pattern in the plane of the aerial's magnetic field. Things to note from these two diagrams are that:

(a) in the case of Fig. 13.20(a) minimum radiation occurs along the axis of the aerial whilst the two zones of maximum radiation are at 90° (i.e. are 'normal to') the dipole elements;
(b) in the case of Fig. 13.20(b) the aerial radiates uniformly in all directions.

Hence, a vertical dipole will have **omnidirectional** characteristics whilst a horizontal dipole will have a **bi-directional** radiation pattern. This is an important point as we shall see later. The combined effect of these two patterns in three-dimensional space will be a doughnut shape.

Example 13.6

Determine the length of a half-wave dipole aerial for use at a frequency of 150 MHz.

Solution

The length of a half-wave dipole for 150 MHz can be determined from, $l = \dfrac{v}{2f}$ where $v = 3 \times 10^8$ m/s and $f = 150 \times 10^6$ Hz.

Hence
$$l = \frac{v}{2f} = \frac{3 \times 10^8}{2 \times 150 \times 10^6} = \frac{3 \times 10^8}{300 \times 10^6}$$
$$= \frac{3 \times 10^6}{3 \times 10^6} = 1 \text{ m}$$

Aerial impedance and radiation resistance

Because voltage and current appear in an aerial (a minute voltage and current in the case of a

receiving antenna and a much larger voltage and current in the case of a transmitting antenna) an aerial is said to have **impedance**. Here it's worth remembering that impedance is a mixture of resistance, R, and reactance, X, both measured in ohms (Ω). Of these two quantities, X varies with frequency whilst R remains constant. This is an important concept because it explains why aerials are often designed for operation over a restricted range of frequencies.

The impedance, Z, of an aerial is the ratio of the voltage, E, across its terminals to the current, I, flowing in it. Hence,

$$Z = \frac{E}{I} \ \Omega$$

You might infer from Fig. 13.19 that the impedance at the centre of the half-wave dipole should be zero. In practice the impedance is usually between $70\,\Omega$ and $75\,\Omega$. Furthermore, at resonance the impedance is purely resistive and contains no reactive component (i.e. inductance and capacitance). In this case X is negligible compared with R. It is also worth noting that the d.c. resistance (or **ohmic resistance**) of an aerial is usually very small in comparison with its impedance and so it may be ignored. Ignoring the d.c. resistance of the aerial, the impedance of an antenna may be regarded as its **radiation resistance**, R_r (see Fig. 13.21).

Radiation resistance is important because it is through this resistance that electrical power is transformed into radiated electromagnetic energy (in the case of a transmitting aerial) and incident

Figure 13.21 Radiation resistance

Figure 13.22   Equivalent circuit of an aerial

electromagnetic energy is transformed into electrical power (in the case of a receiving aerial).

The equivalent circuit of an aerial is shown in Fig. 13.22. The three series-connected components that make up the aerial impedance are:

(a) the d.c. resistance, $R_{d.c.}$;
(b) the radiation d.c. resistance, $R_{d.c.}$;
(c) the 'off-tune' reactance, X.

Note that when the antenna is operated at a frequency that lies in the centre of its pass-band (i.e. when it is *on-tune*) the off-tune reactance is zero. It is also worth bearing in mind that the radiation resistance of a half-wave dipole varies according to its height above ground. The $70\,\Omega$ to $75\,\Omega$ impedance normally associated with a half-wave dipole is only realized when the aerial is mounted at an elevation of 0.2 wavelengths, or more.

Radiated power and efficiency

In the case of a transmitting aerial, the radiated power, P_r, produced by the antenna is given by:

$$P_r = I_a^2 \times R_r \text{ W}$$

I_a is the aerial current, in amperes, and R_r is the radiation resistance in ohms. In most practical applications it is important to ensure that P_r is maximized and this is achieved by ensuring that R_r is much larger than the d.c. resistance of the antenna elements.

The efficiency of an antenna is given by the relationship:

$$\text{Radiation efficiency} = \frac{P_r}{P_r + P_{loss}} \times 100\%$$

where P_{loss} is the power dissipated in the d.c. resistance present. At this point it is worth stating

that, whilst efficiency is vitally important in the case of a transmitting aerial it is generally unimportant in the case of a receiving aerial. This explains why a random length of wire can make a good receiving aerial but not a good transmitting antenna!

Example 13.7

An HF transmitting aerial has a radiation resistance of $12\,\Omega$. If a current of 0.5 A is supplied to it, determine the radiated power.

Solution

In this case, $I_a = 0.5$ A and $R_r = 12\,\Omega$

Since $P_r = I_a^2 \times R_r$

$$P_r = (0.5)^2 \times 12 = 0.25 \times 12 = 4 \text{ W}$$

Example 13.8

If the aerial in Example 13.7 has a d.c. resistance of $2\,\Omega$, determine the power loss and the radiation efficiency of the aerial.

Solution

From the equivalent circuit shown in Fig. 13.22, the same current flows in the d.c. resistance of the aerial, $R_{d.c.}$, as flows in its radiation resistance, R_r.

Hence $I_a = 0.5$ A and $R_{d.c.} = 2\,\Omega$

Since $P_{loss} = I_a^2 \times R_{d.c.}$

$$P_{loss} = (0.5)^2 \times 2 = 0.25 \times 2 = 0.5 \text{ W}$$

The radiation efficiency of the aerial is given by:

$$\begin{aligned}
\text{Radiation efficiency} &= \frac{P_r}{P_r + P_{loss}} \times 100\% \\
&= \frac{4}{4 + 0.5} \times 100\% \\
&= \frac{4}{4.5} \times 100\% = 89\%
\end{aligned}$$

In this example, more than 10% of the power output is actually wasted!

Aerial gain

The field strength produced by an aerial is proportional to the amount of current flowing in it. However, since different types of aerial produce different values of field strength for the same applied RF power level, we attribute a power gain to the aerial. This power gain is specified in relation to a **reference aerial** and it is usually specified in decibels (dB) – see Appendix 5.

Two types of reference aerial are used, an **isotropic radiator** and a **standard half-wave dipole**. The former type of reference aerial is only a theoretical structure (it is assumed to produce a truly spherical radiation pattern and thus could only be realized in three-dimensional space well away from the earth). The latter type of aerial is a more practical reference since it is reasonably easy to produce a half-wave dipole for comparison purposes.

In order to distinguish between the two types of reference aerial we use subscripts **i** and **d** to denote isotropic and half-wave dipole reference aerials respectively. As an example, an aerial having a gain of $10\,dB_i$ produces ten times power gain when compared with a theoretical isotropic radiator. Similarly, an aerial having a gain of $13\,dB_d$ produces 20 times power gain when compared with a half-wave dipole. Putting this another way, to maintain the same field strength at a given point, you would have to apply $20\,W$ to a half-wave dipole or just $1\,W$ to the aerial in question!

Some typical values of aerial gain are given below:

Application	Gain (dB_d)
Half-wave wire dipole for VHF Band II FM broadcast reception	0
Car roof mounted UHF aerial for private mobile radio	4
Four-element Yagi aerial for digital FM broadcast reception	6
Multi-element Yagi aerial for fringe area Band V TV reception	12
Parabolic reflector antenna for satellite television reception	24
3 m diameter steerable dish for tracking space vehicles at UHF	40

Figure 13.23 shows typical half-wave dipole aerials for domestic VHF Band II FM broadcast reception. The half-wave dipole in Fig. 13.23(a) is horizontally polarized (and therefore has a bi-directional characteristic) whilst the half-wave dipole in Fig. 13.23(b) is vertically polarized (and therefore has an omnidirectional characteristic).

Figure 13.23 Typical dipole aerials for VHF Band II FM reception: (a) horizontally polarized (bi-directional) (b) vertically polarized (omnidirectional)

The Yagi beam aerial

Originally invented by two Japanese engineers, Yagi and Uda, the Yagi aerial has remained extremely popular in a wide variety of applications and, in particular, for fixed domestic FM radio and TV receiving aerials. In order to explain, in simple terms how the Yagi aerial works we shall use a simple light analogy.

An ordinary filament lamp radiates light in all directions. Just like an aerial, the lamp converts electrical energy into electromagnetic energy. The only real difference is that we can *see* the energy that it produces!

The action of the filament lamp is comparable with our fundamental dipole aerial. In the case of the dipole, electromagnetic radiation will occur all around the dipole elements (in three dimensions the radiation pattern will take on a doughnut shape). In the plane that we have shown in Fig. 13.20(a), the directional pattern will be a figure-of-eight that has two lobes of equal size. In order to concentrate the radiation into just one of the radiation lobes we could simply place a reflecting mirror on one side of the filament lamp. The radiation will be reflected (during which the reflected light will undergo a 180° phase change) and this will reinforce the light on one side of the filament lamp. In order to achieve the same effect in our aerial system we need to place a conducting element about one-quarter of a wavelength behind the dipole element. This element is referred to as a **reflector** and it is said to be 'parasitic' (i.e. it is not actually connected to the feeder). The reflector needs to be cut slightly longer than the driven dipole element. The resulting directional pattern will now only have one **major lobe** because the energy radiated will be concentrated into just one half of the figure-of-eight pattern that we started with).

Continuing with our optical analogy, in order to further concentrate the light energy into a narrow beam we can add a lens in front of the lamp. This will have the effect of bending the light emerging from the lamp towards the normal line. In order to achieve the same effect in our aerial system we need to place a conducting element, known as a **director**, on the other side of the dipole and about one-quarter of a wavelength from it. Once again, this element is parasitic but in this case it needs to be cut slightly shorter than the driven dipole element. The resulting directional pattern will now

have a narrower major lobe as the energy becomes concentrated in the normal direction (i.e. at right angles to the dipole elements).

The resulting aerial is known as a three-element Yagi aerial, see Fig. 13.24. If desired, additional directors can be added to further increase the gain of the aerial and reduce the **beamwidth** of the major lobe. A typical three-element horizontally polarized Yagi suitable for VHF Band II broadcast reception is shown in Fig. 13.25.

Further director elements can be added to increase the gain and reduce the **beamwidth** (i.e. the angle between the half-power or $-3\,\text{dB}$ power points on the polar characteristic) of Yagi aerials. Some typical gain and beamwidth figures for Yagi aerials are given below.

Number of elements	Gain (dB$_d$)	Beamwidth
3	4	70°
4	6	60°
8	9	40°
16	11.5	20°
32	13	10°

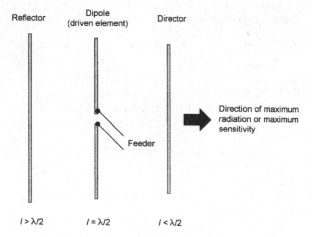

Figure 13.24   A three-element Yagi aerial

Figure 13.25   A typical three-element Yagi aerial for VHF Band II FM reception

Formulae introduced in this chapter

Frequency and wavelength
(page 203)

$$v = f\lambda, f = \frac{v}{\lambda} \text{ and } \lambda = \frac{v}{f}$$

Velocity of propagation
(page 202)

$v = 3 \times 10^8\,\text{m/s}$ for waves travelling in air or space

BFO frequency
(page 204)

$f_{\text{BFO}} = f_{\text{RF}} \pm f_{\text{AF}}$

Local oscillator frequency
(page 208)

$f_{\text{LO}} = f_{\text{RF}} \pm f_{\text{IF}}$

Half-wave dipole aerial
(page 210)

$$l = \frac{\lambda}{2}$$

Aerial impedance
(page 211)

$$Z = \frac{E}{I}$$

Radiated power
(page 210)

$$P_r = I_a^2 \times R_r$$

Radiation efficiency
(page 212)

$$\frac{P_r}{P_r + P_{loss}} \times 100\%$$

Problems

13.1 A broadcast transmitter produces a signal at 190 kHz. What will be the wavelength of the radiated signal?

13.2 What frequency corresponds to the 13 m short wave band?

13.3 A signal in a cable propagates at two-thirds of the speed of light. If an RF signal at 50 MHz is fed to the cable, determine the wavelength in the cable.

13.4 An AM broadcast receiver has an IF of 470 kHz. If the receiver is to be tuned over the medium wave broadcast band from 550 kHz to 1.6 MHz, determine the required local oscillator tuning range.

13.5 The following data was obtained during an experiment on an IF bandpass filter. Plot the frequency response characteristic and use it to determine the IF frequency, bandwidth and Q-factor of the filter.

Frequency (MHz)	10.4	10.5	10.6	10.7	10.8	10.9	11.0
Voltage (V)	0.42	0.52	0.69	1.0	0.67	0.51	0.41

13.6 Refer to the IF amplifier/AM demodulator circuit shown in Fig. 13.17. Identify the component(s) that provide:

(a) a tuned collector load for TR_1;
(b) base bias for TR_1;
(c) coupling of the signal from the IF amplifier stage to the demodulator stage;
(d) a low-pass filter to remove RF signal components at the output of the demodulator;
(e) a volume control;
(f) removing the d.c. level on the signal at the output of the demodulator;
(g) a bypass to the common rail for RF signal that may be present on the supply;
(h) input coupling to the IF amplifier stage.

13.7 A half-wave dipole is to be constructed for a frequency of 50 MHz. Determine the approximate length of the aerial.

13.8 A power of 150 W is applied to a dipole aerial in order to produce a given signal strength at a remote location. What power, applied to a Yagi aerial with a gain of 8 dB_d, would be required to produce the same signal strength?

(Answers to these problems will be found on page 261.)

14

Test equipment and measurements

This chapter is about making practical measurements on real electronic circuits. It describes and explains the use of the basic items of test equipment that you will find in any electronics laboratory or workshop. We begin the chapter by looking at how we use a moving coil meter to measure voltage, current and resistance and then quickly move on to more complex multi-range analogue and digital instruments and the oscilloscope. To help you make use of these test instruments we have included some do's and don'ts. If you intend to become an electronic engineer these will become your 'tools of the trade'. Being able to use them effectively is just the first step on the ladder!

Meters

Straightforward measurements of voltage, current and resistance can provide useful information on the state of almost any circuit. To get the best from a meter it is not only necessary to select an appropriate measurement function and range but also to be aware of the limitations of the instrument and the effect that it might have on the circuit under investigation. When fault finding, it is interpretation that is put on the meter readings rather than the indications themselves.

Figures 14.2(a) and 14.2(b) respectively show the circuit of a simple **voltmeter** and a simple **ammeter**. Each instrument is based on the moving coil indicator shown in Fig. 14.1. The voltmeter consists of a **multiplier** resistor connected in series with the basic moving coil movement whilst the ammeter consists of a **shunt** resistor connected in parallel with the basic moving coil instrument.

When determining the value of multiplier or shunt resistance (R_m and R_s respectively in Fig. 14.2) it is important to remember that the coil of the moving coil meter also has a resistance. We have shown this as a resistor, r, connected with the moving coil in Fig. 14.3. In both cases, the current required to produce full-scale deflection of the meter movement is I_m.

In the voltmeter circuit shown in Fig. 14.3(a):

$$V = I_m R_m + I_m r$$

from which:

$$I_m R_m = V - I_m r$$

Figure 14.1 A moving coil meter movement

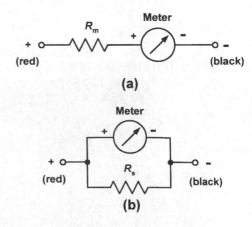

Figure 14.2 A moving coil meter connected as (a) a voltmeter and (b) an ammeter

Figure 14.3 Circuit for determining the values of (a) a multiplier resistor and (b) a shunt resistor

Thus:

$$R_m = \frac{V - I_m r}{I_m}$$

In the ammeter circuit shown in Fig. 14.3(b):

$$(I - I_m)R_s = I_m r$$

from which:

$$R_s = \frac{I_m r}{I - I_m}$$

Example 14.1

A moving coil meter has a full-scale deflection current of 1 mA. If the meter coil has a resistance of 500 Ω, determine the value of multiplier resistor if the meter is to be used as a voltmeter reading 0 to 5 V.

Solution

Using $R_m = \dfrac{V - I_m r}{I_m}$ gives:

$$R_m = \frac{5 - (1 \times 10^{-3} \times 500)}{1 \times 10^{-3}} = \frac{5 - 0.5}{1 \times 10^{-3}}$$

$$= 4.5 \times 10^3 = 4.5\,k\Omega$$

Example 14.2

A moving coil meter has a full-scale deflection current of 10 mA. If the meter coil has a resistance of 40 Ω, determine the value of shunt resistor if the meter is to be used as an ammeter reading 0 to 100 mA.

Solution

Using $R_s = \dfrac{I_m r}{I - I_m}$ gives:

$$R_s = \frac{10 \times 10^{-3} \times 40}{100 \times 10^{-3} - 10 \times 10^{-3}} = \frac{400 \times 10^{-3}}{90 \times 10^{-3}}$$

$$= \frac{400}{90} = 4.44\,\Omega$$

Ohmmeters

The circuit of a simple ohmmeter is shown in Fig. 14.4. The battery is used to supply a current that will flow in the unknown resistor, R_x, which is indicated on the moving coil meter. Before use, the variable resistor, RV, must be adjusted in order to produce full-scale deflection (corresponding to zero on the ohms scale). Zero resistance thus corresponds to maximum indication. Infinite resistance (i.e. when the two terminals are left open-circuit) corresponds to minimum indication. The ohms scale is thus reversed when compared with a voltage or current scale. The scale is also non-linear, as shown in Fig. 14.5.

Multi-range meters

For practical measurements on electronic circuits it is often convenient to combine the functions of a

Figure 14.4 A simple ohmmeter

Figure 14.5 A typical ohmmeter scale

Figure 14.6 A typical seven-segment display

voltmeter, ammeter and ohmmeter into a single instrument (known as a multi-range meter or simply a **multimeter**). In a conventional multimeter as many as eight or nine measuring functions may be provided with up to six or eight ranges for each measuring function. Besides the normal voltage, current and resistance functions, some meters also include facilities for checking transistors and measuring capacitance. Most multi-range meters normally operate from internal batteries and thus they are independent of the mains supply. This leads to a high degree of portability which can be all-important when measurements are to be made away from a laboratory or workshop.

Displays

Analogue instruments employ conventional moving coil meters and the display takes the form of a pointer moving across a calibrated scale. This arrangement is not so convenient to use as that employed in digital instruments because the position of the pointer is rarely exact and may require interpolation. Analogue instruments do, however, offer some advantages not the least of which lies in the fact that it is very easy to make adjustments to a circuit whilst observing the relative direction of the pointer; a movement in one direction representing an increase and in the other a decrease. Despite this, the principal disadvantage of many analogue meters is the rather cramped, and sometimes confusing, scale calibration. To determine the exact reading requires first an estimation of the pointer's position and then the application of some mental arithmetic based on the range switch setting.

Digital meters, on the other hand, are usually extremely easy to read and have displays that are clear, unambiguous, and capable of providing a very high resolution. It is thus possible to distinguish between readings that are very close. This is just not possible with an analogue instrument. A typical seven-segment display used

in digital readout equipment is shown in Fig. 14.6. The type of display used in digital multi-range meters is either the liquid crystal display (LCD) or the light emitting diode (LED). The former type requires very little electrical power and thus is to be preferred on the grounds of low battery consumption. LCD displays are, however, somewhat difficult to read under certain light conditions and, furthermore, the display response can be rather slow. LED displays can be extremely bright but unfortunately consume considerable power and this makes them unsuitable for use in battery-powered portable instruments.

Loading

Another very significant difference between analogue and digital instruments is the input resistance that they present to the circuit under investigation when taking voltage measurements. The resistance of a reasonable quality analogue multi-range meter can be as low as 50 kΩ on the 2.5 V range. With a digital instrument, on the other hand, the input resistance is typically 10 MΩ on the 2 V range. The digital instrument is thus to be preferred when accurate readings are to be taken. This is particularly important when measurements are to be made on high impedance circuits, as illustrated by the following.

Two multi-range meters are used to measure the voltage produced by the two potential dividers shown in Fig. 14.7. One of the meters is an analogue type having an internal resistance of 10 kΩ on the 10 V range. The other is a digital type that has the much higher internal resistance of 10 MΩ. The two potential dividers each consist of resistors of identical value. However, the potential divider of Fig. 14.7(b) has a much lower resistance than that of Fig. 14.7(a). In both cases the 'true' voltage produced by the potential divider should be half the supply, i.e. exactly 5 V.

Figure 14.7 Examples of voltmeter loading: (a) a low resistance circuit (b) a high resistance circuit

The actual readings obtained from the instruments are as follows:

Meter	High resistance circuit (Fig. 14.7(a))	Low resistance circuit (Fig. 14.7(b))
Analogue type (100 kΩ)	4.9 V	3.3 V
Digital type (10 MΩ)	4.99 V	4.97 V

The large difference in the case of Fig. 14.7(b) illustrates the effect of voltmeter **loading** on a high resistance circuit. An appreciable current is drawn away from the circuit into the measuring instrument. Clearly this is a very undesirable effect!

Sensitivity

The sensitivity of an analogue meter may be expressed in several ways. One is to specify the basic **full-scale deflection** (f.s.d.) current of the moving coil meter. This is typically $50\,\mu\text{A}$ or less. An alternative method is that of quoting an **ohms-per-volt** rating. This is, in effect, the resistance presented by the meter when switched to the 1 V range.

The **ohms-per-volt** rating is inversely proportional to the basic full-scale sensitivity of the meter movement and, to determine the resistance of a meter on a particular voltage range, it is only necessary to multiply the range setting by the 'ohms-per-volt' rating. The following table shows how meter f.s.d. and ohms-per-volt are related:

Meter f.s.d.	Ohms-per-volt (kΩ/V)
$10\,\mu\text{A}$	100
$20\,\mu\text{A}$	50
$50\,\mu\text{A}$	20
$100\,\mu\text{A}$	10
$200\,\mu\text{A}$	5
$500\,\mu\text{A}$	2
$1\,\text{mA}$	1

From the above we can conclude that:

$$\text{Meter f.s.d.} = \frac{1}{\text{Ohms-per-volt}}$$

or

$$\text{Ohms-per-volt} = \frac{1}{\text{Meter f.s.d.}}$$

Example 14.3

A meter has a full-scale deflection of $40\,\mu\text{A}$. What will its ohms-per-volt rating be?

Solution

$$\text{Ohms-per-volt} = \frac{1}{\text{Meter f.s.d.}} = \frac{1}{40 \times 10^{-6}}$$
$$= 25 \times 10^3 = 25\,\text{k}\Omega$$

Example 14.4

A $20\,\text{k}\Omega/\text{V}$ meter is switched to the 10 V range. What current will flow in the meter when it is connected to a 6 V source?

Solution

The resistance of the meter will be given by:

$R_m = 10 \times 20\,k\Omega = 200\,k\Omega$

The current flowing in the meter will thus be given by:

$$I_m = \frac{6\,V}{R_m} = \frac{6}{200 \times 10^3} = 30 \times 10^{-6} = 30\,\mu A$$

Digital multi-range meters

Low-cost digital multi-range meters have been made possible by the advent of mass-produced LSI devices and liquid crystal displays. A $3\frac{1}{2}$-digit display is the norm and this consists of three full digits that can display '0' to '9' and a fourth (most significant) digit which can only display '1'. Thus, the maximum display indication, ignoring the range switching and decimal point, is 1999; anything greater over-ranges the display.

The **resolution** of the instrument is the lowest increment that can be displayed and this would normally be an increase or decrease of one unit in the last (least significant) digit. The **sensitivity** of a digital instrument is generally defined as the smallest increment that can be displayed on the lowest (most sensitive) range. Sensitivity and resolution are thus not quite the same. To put this into context, consider the following example.

A digital multi-range meter has a $3\frac{1}{2}$-digit display. When switched to the 2 V range, the maximum indication would be 1.999 V and any input of 2 V, or greater, would produce an over-range indication. On the 2 V range, the instrument has a resolution of 0.001 V (or 1 mV). The lowest range of the instrument is 200 mV (corresponding to a maximum display of 199.9 mV) and thus the meter has a sensitivity of 0.1 mV (or 100 μV).

Nearly all digital meters have automatic zero and polarity indicating facilities and some also have **autoranging**. This feature, which is only found in the more expensive instruments, automatically changes the range setting so that maximum resolution is obtained without over-ranging. There is thus no need for manual operation of the range switch once the indicating mode has been selected. This is an extremely useful facility since it frees the user from the need to make repeated adjustments to the range switch while measurements are being made.

Example 14.3

A digital multi-range meter has a $4\frac{1}{2}$-digit display. When switched to the 200 V range, determine:

(a) the maximum indication that will appear on the display;
(b) the resolution of the instrument.

Solution

(a) The maximum indication that will appear on the display is 199.99 V.
(b) The resolution of the instrument will be 0.01 V or 10 mV.

Using an analogue multi-range meter

Figure 14.8 shows the controls and display provided by a simple analogue multi-range meter. The range selector allows you to select from a total of 20 ranges and six measurement functions. These functions are:

D.c. voltage (DC, V)
D.c. current (DC, mA)
A.c. voltage (AC, V)
Resistance (OHM)
Continuity test (BUZZ)
Battery check (BAT)

D.c. voltage measurement

Figure 14.9 shows how to make d.c. voltage measurements. In both cases, the red and black test leads are connected to the '+' and '−' sockets respectively. In Fig. 14.9, the range selector is set to DCV, 50 V. The pointer is reading just less than 45 on the range that has 50 as its full-scale indication (note that there are three calibrated voltage scales with maximum indications of 10 V, 50 V and 250 V respectively. The reading indicated is approximately 45 V.

D.c. current measurement

Figure 14.10 shows how to make a d.c. current measurement. Once again, the red and black test leads are connected to the '+' and '−' sockets respectively. The range selector is set to DC, 5 mA. In Fig. 14.10, the pointer is reading between 35 and 40 (and is actually a little closer to 35 than it is to 40) on the range that has 50 as its full-scale indication. The actual reading indicated is thus approximately 37 mA.

Figure 14.8 Analogue multi-range meter display and controls

Figure 14.9 Analogue multi-range meter set to DC, 50 V range

Figure 14.10   Analogue multi-range meter set to DC, 5 mA range

Figure 14.11   Analogue multi-range meter set to DC, 10 A range

d.c. high current measurement

In common with many simple multi-range meters, both analogue and digital, the high current range (e.g. 10 A) is not only selected using the range selector switch but a separate input connection must also be made. The reason for this is simply that the range switch and associated wiring is not designed to carry a high current. Instead, the high current shunt is terminated separately at its own '10 A' socket.

Figure 14.11 shows the connections and range selector settings to permit high current d.c. measurement. The range selector is set to DC 10 A and the red and black test leads are connected to '10 A' and '−' respectively. The pointer is reading mid-way between 9 and 10 on the range that has 10 as its full-scale indication. The actual reading indicated is thus 9.5 A.

a.c. voltage measurement

Figure 14.12 shows how to make a.c. voltage measurements. Once again, the red and black test leads are connected to the '+' and '−' sockets respectively. In Fig. 14.12, the range selector is set to AC, 10 V. The pointer is reading mid-way between 2 and 3 and the indicated reading is approximately 2.5 V.

Output level measurement

Figure 14.13 shows how to make output level measurements. The red and black test leads are respectively connected to 'OUT' and '−' respectively. The range selector is set to AC, 10V (note that the output level facility is actually based on a.c. voltage measurement).

Output level indications are indicated in decibels (dB) where 0 dB (the 'reference' level) corresponds to a power level of 1 mW in a resistance of 600 Ω. The pointer is reading mid-way between +18 and +20 on the dB scale and the indicated reading is +19 dB.

Resistance

Figure 14.14 shows how to make resistance measurements. In all three cases, the red and black test leads are connected to the '+' and '−'

Figure 14.12   Analogue multi-range meter set to AC, 10 V range

Figure 14.13 Analogue multi-range meter set to dB (output level) range

sockets respectively. Before making any measurements it is absolutely essential to zero the meter. This is achieved by shorting the test leads together and adjusting the ZERO ADJ control until the meter reads full scale (i.e. zero on the ohms scale). In Fig. 14.14, the range selector is set to OHM, × 1. The pointer is reading mid-way between 0 and 10 and the resistance indicated is approximately 5 Ω.

DO's and DON'Ts of using an analogue multi-range meter

DO ensure that you have selected the correct range and measuring function before attempting to connect the meter into a circuit.

DO select a higher range than expected and then progressively increase the sensitivity as necessary to obtain a meaningful indication.

DO remember to zero on the ohms range before measuring resistance.

DO switch the meter to the 'off' position (if one is available) before attempting to transport the meter.

DO check and, if necessary, replace the internal batteries regularly.

DO use properly insulated test leads and prods.

DON'T attempt to measure resistance in a circuit that has the power applied to it.

DON'T rely on voltage readings made on high impedance circuits (the meter's own internal resistance may have a significant effect on the voltages that you measure).

DON'T rely on voltage and current readings made on circuits where high frequency signals may be present (in such cases an analogue meter may produce readings that are wildly inaccurate or misleading).

DON'T subject the instrument to excessive mechanical shock or vibration (this can damage the sensitive meter movement).

Using a digital multi-range meter

Digital multi-range meters offer a number of significant advantages when compared with their

Figure 14.14   Analogue multi-range meter set to OHM, ×1 range

more humble analogue counterparts. The display fitted to a digital multi-range meter usually consists of a $3\frac{1}{2}$-digit seven-segment display – the $\frac{1}{2}$ simply indicates that the first digit is either blank (zero) or 1. Consequently, the maximum indication on the 2 V range will be 1.999 V and this shows that the instrument is capable of offering a resolution of 1 mV on the 2 V range. The resolution obtained from a comparable analogue meter would be of the order of 50 mV, or so, and thus the digital instrument provides a resolution that is many times greater than its analogue counterpart.

Figure 14.15 shows the controls and display provided by a simple analogue multi-range meter. The mode switch and range selector allow you to select from a total of 20 ranges and eight measurement functions. These functions are:

D.c. voltage (DC, V)
D.c. current (DC, A)
A.c. voltage (AC, V)
A.c. current (AC, A)
Resistance (OHM)
Capacitance (CAP)
Continuity test (buzzer)
Transistor current gain (h_{fe})

d.c. voltage measurement

Figure 14.16 shows how to make d.c. voltage measurements using a digital multi-range meter. The red and black test leads are connected to the 'V-OHM' and 'COM' sockets respectively. In Fig. 14.16, the mode switch and range selector is set to DCV, 200 V, and the display indicates a reading of 124.5 V.

d.c. current measurement

Figure 14.17 shows how to make a d.c. current measurement. Here, the red and black test leads are connected to the 'mA' and 'COM' sockets respectively. The mode switch and range selectors are set to DC, 200 mA, and the display indicates a reading of 85.9 mA.

d.c. high current measurement

In common with simple analogue multi-range meters, the meter uses a shunt which is directly connected to a separate '10 A' terminal. Figure 14.18 shows the connections, mode switch and range selector settings to permit high current d.c.

Figure 14.15 Digital multi-range meter display and controls

measurement. The mode switch and range selectors are set to DC, 2000 mA (2 A) and the red and black test leads are connected to '10 A' and 'COM' respectively. The display indicates a reading of 2.99 A.

a.c. voltage measurement

Figure 14.19 shows how to make a.c. voltage measurements. Once again, the red and black test leads are connected to the 'V-OHM' and 'COM' sockets respectively. In Fig. 14.19, the mode switch and range selectors are set to AC, 10 V, and the display indicates a reading of 1.736 V.

Resistance measurement

Figure 14.20 shows how to make resistance measurements. As before, the red and black test leads are connected to 'V-OHM' and 'COM' respectively. In Fig. 14.18, the mode switch and range selectors are set to OHM, 200 ohm, and the meter indicates a reading of 55.8 Ω. Note that it is not necessary to 'zero' the meter by shorting the test probes together before taking any measurements (as would be the case with an analogue meter).

Capacitance measurement

Many modern digital multi-range meters incorporate a capacitance measuring although this may be limited to just one or two ranges. Figure 14.21 shows how to carry out a capacitance measurement. The capacitor on test is inserted into the two-way connector marked 'CAP' whilst the mode switch and range selector controls are set to DC, 2000 pF. The display indication shown in Fig. 14.21 corresponds to a capacitance of 329 pF.

Figure 14.16   Digital multi-range meter set to DC, 200 V range

Figure 14.17   Digital multi-range meter set to DC, 200 mA range

Figure 14.18   Digital multi-range meter set to DC, 10 A range

Figure 14.19   Digital multi-range meter set to AC, 2 V range

Figure 14.20 Digital multi-range meter set to OHM, 200 Ω range

Figure 14.21 Digital multi-range meter set to capacitance, 2000 pF range

Transistor current gain (h_{fe}) measurement

Many modern digital multi-range meters also provide some (fairly basic) facilities for checking transistors. Figure 14.22 shows how to measure the current gain (h_{fe}) of an NPN transistor (see page 93). The transistor is inserted into the three-way connector marked 'EBC', taking care to ensure that the emitter lead is connected to 'E', the base lead to 'B', and the collector lead to 'C'. The mode switch and range selector controls are set to DC, NPN respectively. The display indication in Fig. 14.22 shows that the device has a current gain of 93. This means that, for the device in question, the ratio of collector current (I_C) to base current (I_B) is 93.

Connecting a meter into a circuit

Figure 14.23 shows how a meter is connected to read the voltages and currents present at various points in a simple transistor audio amplifier.

(a) At A, the supply current is measured by removing the **supply rail fuse** and inserting the meter in its place. A suitable meter range is 2 A, d.c.

(b) At B, the collector current for TR4 is measured by removing the **service link** and inserting the meter in its place. A suitable meter range is 2 A, d.c.

(c) At C, the emitter current for TR4 can be measured by connecting the meter across the emitter resistor, R11, and measuring the **voltage drop** across it. The emitter current can then be calculated using Ohm's law (see page 6). This can be much quicker than disconnecting the emitter lead and inserting the meter switched to the current range. A suitable range for the meter is 2 V, d.c. (note that 2 V is dropped across 1 Ω when a current of 2 A is flowing through it).

(d) At D, the voltage at the junction of R11 and R12 is measured by connecting the meter between the junction of R11 and R12 and common/ground. A suitable meter range is 20 V, d.c.

Figure 14.22   Digital multi-range meter set to transistor current gain, NPN range

Figure 14.23 Connecting a meter to display currents and voltages in a simple audio amplifier

(e) At E, the base–emitter voltage for TR2 is measured by connecting the meter with its positive lead to the base of TR2 and its negative lead to the emitter of TR2. A suitable meter range is 2 V, d.c.

DO's and DON'Ts of using a digital multi-range meter

DO ensure that you have selected the correct range and measuring function before attempting to connect the meter into a circuit.

DO select a higher range than expected and then progressively increase the sensitivity as necessary to obtain a meaningful indication.

DO switch the meter to the 'off' position in order to conserve battery life when the instrument is not being used.

DO check and, if necessary, replace the internal battery regularly.

DO use properly insulated test leads and prods.

DO check that a suitably rated fuse is used in conjunction with the current ranges.

DON'T attempt to measure resistance in a circuit that has the power applied to it.

DON'T rely on voltage and current readings made on circuits where high frequency signals may be present (as with analogue instruments, digital meters may produce readings that are wildly inaccurate or misleading in such circumstances).

DON'T rely on measurements made when voltage/current is changing or when a significant amount of a.c. may be present superimposed on a d.c. level.

The oscilloscope

An oscilloscope is an extremely comprehensive and versatile item of test equipment which can be used in a variety of measuring applications, the most important of which is the display of time-related voltage waveforms.

The oscilloscope display is provided by a **cathode ray tube** (CRT) that has a typical screen area of 8 cm × 10 cm. The CRT is fitted with a **graticule** that may either be integral with the tube face or a separate translucent sheet. The graticule is usually ruled with a 1 cm grid to which further bold lines may be added to mark the major axes on the central viewing area. Accurate voltage and time measurements may be made with reference to the graticule, applying a scale factor derived from the appropriate range switch.

A word of caution is appropriate at this stage, however. Before taking meaningful measurements from the CRT screen it is absolutely essential to ensure that the front panel variable controls are set in the **calibrate** (CAL) position. Results will almost certainly be inaccurate if this is not the case!

The use of the graticule is illustrated by the following example.

An oscilloscope screen is depicted in Fig. 14.24. This diagram is reproduced actual size and the fine graticule markings are shown every 2 mm along the central vertical and horizontal axes. The oscilloscope is operated with all relevant controls in the 'CAL' position. The timebase (horizontal deflection) is

switched to the 1 ms/cm range and the vertical attenuator (vertical deflection) is switched to the 1 V/cm range. The overall height of the trace is 4 cm and thus the peak–peak voltage is $4 \times 1\,V = 4\,V$. Similarly, the time for one complete cycle (period) is $2.5 \times 1\,ms = 2.5\,ms$. One further important piece of information is the shape of the waveform that, in this case, is sinusoidal (see page 67).

The front panel layout for a typical general-purpose two-channel oscilloscope is shown in Figs 14.25 and 14.26. The following controls and adjustments are provided in the table in page 233.

Using an oscilloscope

An oscilloscope can provide a great deal of information about what is going on in a circuit. In effect, it allows you to 'see' into the circuit, displaying waveforms that correspond to the signals that are present. The procedure and adjustments differ according to the type of waveform being investigated and whether the oscilloscope is being used to display a single waveform (i.e. single-channel operation) or whether it is being used to display two waveforms simultaneously (i.e. dual-channel operation).

Sinusoidal waveforms (single-channel operation)

The procedure for displaying a repetitive sine wave waveform is shown in Fig. 14.27. The signal is connected to the Channel 1 input (with 'AC' input selected) and the mode switch in the Channel 1 position. 'Channel 1' must be selected as the trigger source and the trigger level control adjusted for a stable display. Where accurate measurements are required it is essential to ensure that the 'Cal' position is selected for both the variable gain and time controls.

Square waveforms (single-channel operation)

The procedure for displaying a repetitive square waveform is shown in Fig. 14.28. Once again, the signal is connected to the Channel 1 input (but this time with 'DC' input selected) and the mode switch in the Channel 1 position. 'Channel 1' must be selected as the trigger source and the trigger level control adjusted for a stable display (which can be triggered on the positive- or negative-going edge of the waveform according to the setting of the trigger polarity button). Any d.c. level present

2.5cm = 2.5ms

4cm = 4V

Timebase: 1ms/cm

Vertical attenuator: 1V/cm

Figure 14.24   Using an oscilloscope graticule

Control	Adjustment
Cathode ray tube display	
Focus	Provides a correctly focused display on the CRT screen.
Intensity	Adjusts the brightness of the display.
Astigmatism	Provides a uniformly defined display over the entire screen area and in both x- and y-directions. The control is normally used in conjunction with the focus and intensity controls.
Trace rotation	Permits accurate alignment of the display with respect to the graticule.
Scale illumination	Controls the brightness of the graticule lines.
Horizontal deflection system	
Timebase (time/cm)	Adjusts the timebase range and sets the horizontal time scale. Usually this control takes the form of a multi-position rotary switch and an additional continuously variable control is often provided. The 'CAL' position is usually at one, or other, extreme setting of this control.
Stability	Adjusts the timebase so that a stable displayed waveform is obtained.
Trigger level	Selects the particular level on the triggering signal at which the timebase sweep commences.
Trigger slope	This usually takes the form of a switch that determines whether triggering occurs on the positive- or negative-going edge of the triggering signal.
Trigger source	This switch allows selection of one of several waveforms for use as the timebase trigger. The options usually include an internal signal derived from the vertical amplifier, a 50 Hz signal derived from the supply mains, and a signal which may be applied to an external trigger input.
Horizontal position	Positions the display along the horizontal axis of the CRT.
Vertical deflection system	
Vertical attenuator (V/cm)	Adjusts the magnitude of the signal attenuator (V/cm) displayed and sets the vertical voltage scale. This control is invariably a multi-position rotary switch; however, an additional variable gain control is sometimes also provided. Often this control is concentric with the main control and the 'CAL' position is usually at one, or other, extreme setting of the control.
Vertical position	Positions the display along the vertical axis of the CRT.
a.c.-d.c.-ground	Normally an oscilloscope employs d.c. coupling throughout the vertical amplifier; hence a shift along the vertical axis will occur whenever a direct voltage is present at the input. When investigating waveforms in a circuit one often encounters a.c. superimposed on d.c. levels; the latter may be removed by inserting a capacitor in series with the signal. With the a.c.-d.c.-ground switch in the a.c. position a capacitor is inserted in the input lead, whereas in the d.c. position the capacitor is shorted. If ground is selected, the vertical input is taken to common (0 V) and the oscilloscope input is left floating. This last facility is useful in allowing the accurate positioning of the vertical position control along the central axis. The switch may then be set to d.c. and the magnitude of any d.c. level present at the input may be easily measured by examining the shift along the vertical axis.
Chopped-alternate	This control, which is only used in dual beam oscilloscopes, provides selection of the beam splitting mode. In the chopped position, the trace displays a small portion of one vertical channel waveform followed by an equally small portion of the other. The traces are, in effect, sampled at a relatively fast rate, the result being two apparently continuous displays. In the alternate position, a complete horizontal sweep is devoted to each channel alternately.

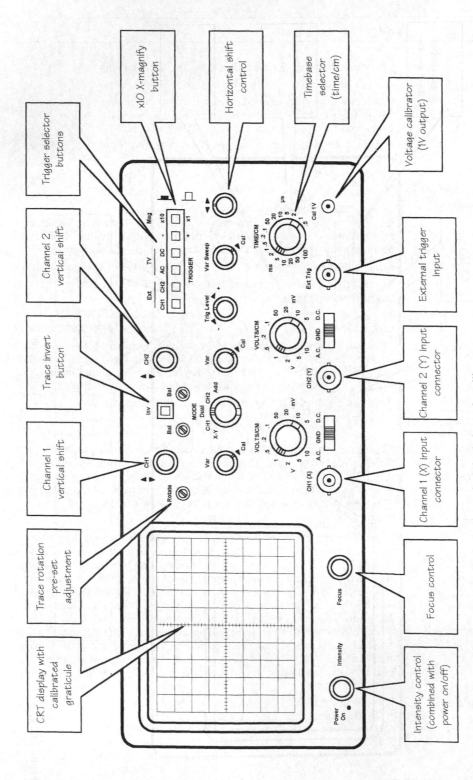

Figure 14.25 Front panel controls and display on a typical dual-channel oscilloscope

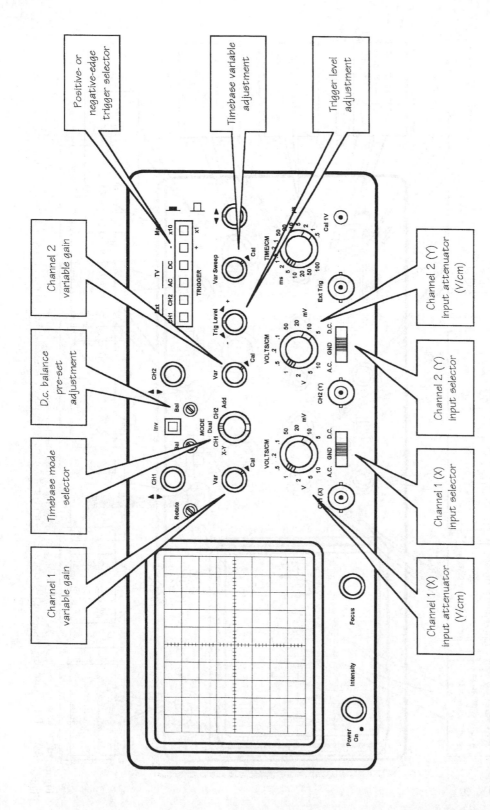

Figure 14.26 Front panel controls and display on a typical dual-channel oscilloscope

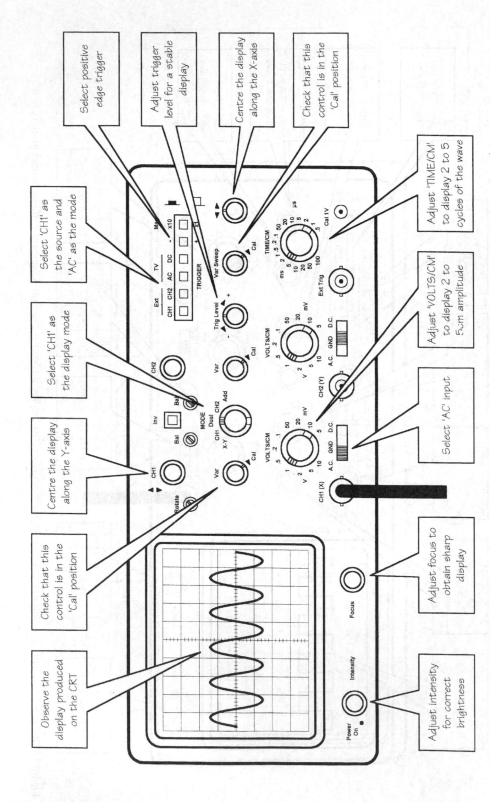

Figure 14.27 Procedure for adjusting the controls of an oscilloscope to display a sinusoidal waveform (single-channel mode)

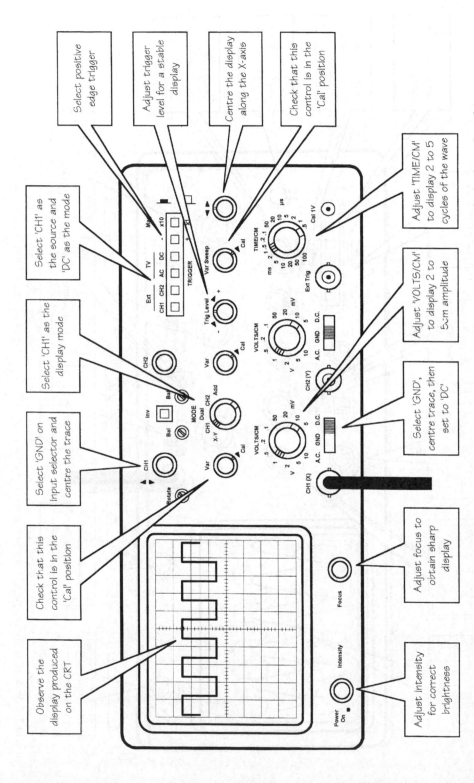

Figure 14.28 Procedure for adjusting the controls of an oscilloscope to display a repetitive square waveform (single-channel mode)

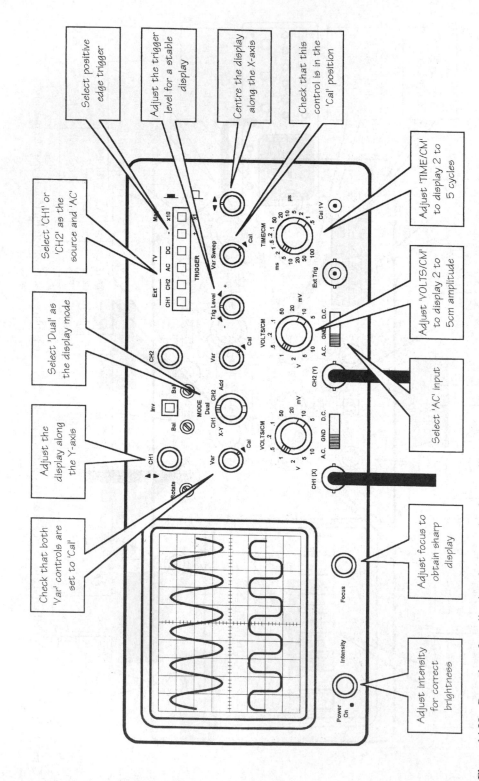

Figure 14.29 Procedure for adjusting the controls of an oscilloscope to display two waveforms simultaneously (dual-channel mode)

Figure 14.30 Connecting an oscilloscope to display waveforms in a simple audio amplifier

on the input can be measured from the offset produced on the Y-axis. To do this, you must first select 'GND' on the input selector then centre the trace along the Y-axis before switching to 'DC' and noting how far up or down the trace moves (above or below 0 V). This may sound a little difficult but it is actually quite easy to do! The same technique can be used for measuring any d.c. offset present on a sinusoidal signal.

Sine/square waveforms (dual-channel operation)

The procedure for displaying two waveforms (either sine or square or any other repetitive signal) simultaneously is shown in Fig. 14.29. The two signals are connected to their respective inputs (Channel 1 and Channel 2) and the mode

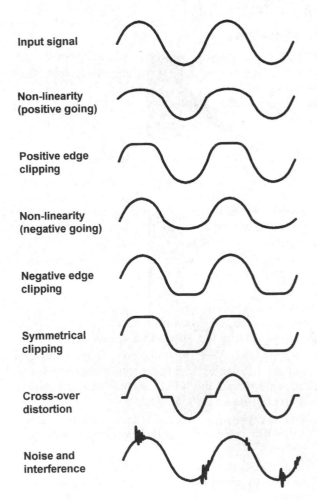

Input signal

Non-linearity (positive going)

Positive edge clipping

Non-linearity (negative going)

Negative edge clipping

Symmetrical clipping

Cross-over distortion

Noise and interference

Figure 14.31 Typical waveforms produced by different types of distortion

switch set to the 'Dual' position. The oscilloscope can be triggered by either of the signals (Channel 1 or Channel 2) as desired. Once again, the display can be triggered on the positive- or negative-going edge of the waveform depending upon the setting of the trigger polarity button. Dual-channel operation can be invaluable when it is necessary to compare two waveforms, e.g. the input and output waveforms of an amplifier.

Connecting an oscilloscope into a circuit

Figure 14.30 shows how an oscilloscope can be connected to display the waveforms present at various points in a simple transistor audio amplifier. To reduce the likelihood of picking up hum and noise, the input to the oscilloscope is via a screened lead fitted with a probe. The ground (outer screen) is connected to the common 0 V rail whilst the probe is simply moved around the circuit from point to point. Note that, because of the ground connection to the oscilloscope it is *not* usually possible to display a waveform that appears 'across' a component (e.g. between the base and emitter of a transistor). For this reason, waveforms are nearly always displayed relative to ground (or common).

Checking distortion

Oscilloscopes are frequently used to investigate **distortion** in amplifiers and other electronic systems. Different forms of distortion have a different effect on a waveform and thus it is possible to determine which type of distortion is present. A 'pure' sine wave is used as an input signal and the output is then displayed on the oscilloscope. Figure 14.31 shows waveforms that correspond to the most common forms of distortion.

Checking frequency response

An oscilloscope can also be used to provide a rapid assessment of the frequency response of an amplifier or other electronic system. Instead of using a sine wave as an input signal a square wave input is used. A different frequency response produces a different effect on a waveform and thus it is possible to assess whether the frequency response is good or poor (a perfect square wave

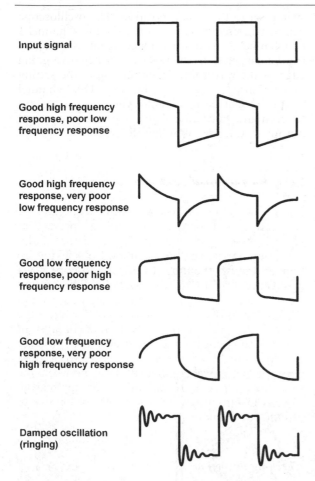

Input signal	
Good high frequency response, poor low frequency response	
Good high frequency response, very poor low frequency response	
Good low frequency response, poor high frequency response	
Good low frequency response, very poor high frequency response	
Damped oscillation (ringing)	

Figure 14.32   Using a square waveform to rapidly assess frequency response

output corresponds to a perfect frequency response). Figure 14.32 shows waveforms that correspond to different frequency response characteristics.

Measuring pulse parameters

When dealing with rectangular waveforms and pulses it is often necessary to be able to use an oscilloscope to measure parameters such as:

Periodic time, t

This is the time measured at the 50% amplitude points for one complete cycle of a repetitive pulse waveform. The periodic time is sometimes referred to as the **period** (see page 67).

On time, t_{on}

This is the time for which the pulse amplitude exceeds 50% of its amplitude. The on time is sometimes referred to as the **high time**.

Off time, t_{off}

This is the time for which the pulse amplitude falls below 50% of its amplitude. The off time is sometimes referred to as the **low time**.

Rise time, t_{rise}

This is the time measured between the 10% and 90% points on the rising (or **positive-going**) edge of the pulse.

Fall time, t_{fall}

This is the time measured between the 90% and 10% points on the falling (or **negative-going**) edge of the pulse.

These **pulse parameters** are shown in Fig. 14.33.

DO's and DON'Ts of using an oscilloscope

DO ensure that the vertical gain and variable timebase controls are set to the calibrate (CAL) positions before attempting to make accurate measurements based on the attenuator/timebase settings and graticule.

DO check that you have the correct trigger source selected for the type of waveform under investigation.

DO remember to align the trace with the X-axis of the graticule with the input selector set to 'GND' prior to taking d.c. offset measurements.

DO make use of the built-in calibrator facility (where available) to check the accuracy of the attenuator and the 'CAL' setting of the variable gain control.

DO use properly screened leads and a suitable probe for connecting to the circuit under investigation.

DO check that you have made a proper connection to ground or 0 V prior to taking measurements.

DON'T leave the intensity control set at a high level for any length of time.

Figure 14.33 Pulse parameters

Figure 14.34 See Problem 14.6

Figure 14.35   See Problem 14.7

Figure 14.36   See Problem 14.8

DON'T leave a bright spot on the display for even the shortest time (this may burn the phosphor coating of the screen).

DON'T rely on voltage measurements on circuits where high frequency signals may be outside the bandwidth of the oscilloscope.

Formulae introduced in this chapter

Voltmeter multiplier resistor
(page 217)

$$R_\mathrm{m} = \frac{V - I_\mathrm{m}r}{I_\mathrm{m}}$$

Ammeter shunt resistor
(page 217)

$$R_\mathrm{s} = \frac{I_\mathrm{m}r}{I - I_\mathrm{m}}$$

Ohms-per-volt
(page 219)

$$\text{Ohms-per-volt} = \frac{1}{\text{Meter f.s.d.}}$$

Problems

14.1 A moving coil meter has a full-scale deflection current of 500 µA. If the meter coil has a resistance of 400 Ω, determine the value of multiplier resistor if the meter is to be used as a voltmeter reading 0 to 10 V.

14.2 A moving coil meter has a full-scale deflection current of 5 mA. If the meter coil has a resistance of 100 Ω, determine the value of shunt resistor if the meter is to be used as an ammeter reading 0 to 50 mA.

14.3 A meter has a full-scale deflection of 60 µA. What will its ohms-per-volt rating be?

14.4 A 50 kΩ/V meter is switched to the 3 V range. What current will flow in the meter when it is connected to a 2 V source?

14.5 A digital multi-range meter has a $3\frac{1}{2}$-digit display. When switched to the 20 V range, determine:

(a) the maximum indication that will appear on the display;
(b) the resolution of the instrument.

14.6 Determine the reading on the multi-range meter shown in Fig. 14.34.

14.7 Determine the reading on the multi-range meter shown in Fig. 14.35.

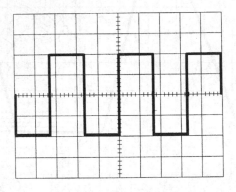

(a)   **Timebase: 1 ms/cm**
      Y-attenuator: 1 V/cm

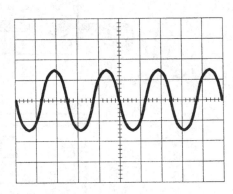

(b)   **Timebase: 50 ns/cm**
      Y-attenuator: 50 mV/cm

(c)   **Timebase: 100 ms/cm**
      Y-attenuator: 50 V/cm

Figure 14.37   See Problem 14.9

Timebase: 1μs/cm

Y-attenuator: 1V/cm

Figure 14.38   See Problem 14.10

14.8   Determine the reading on the multi-range meter shown in Fig. 14.36.

14.9   Figure 14.37 shows the display of an oscilloscope in which the graticule is divided into squares of 1 cm. Determine the amplitude and period of each waveform.

14.10 Figure 14.38 shows the display of an oscilloscope in which the graticule is divided into squares of 1 cm. For the square wave pulse shown determine:

(a) the periodic time;
(b) the high time;
(c) the low time;
(d) the rise time;
(e) the fall time;
(f) the pulse amplitude.

(Answers to these problems will be found on page 262.)

Appendix 1

Student assignments

The 21 student assignments provided here have been designed to support the topics introduced in this book. In conjunction with a taught course, they can also be used to satisfy the assessment requirements of several awarding bodies. Please note that the assignments are not exhaustive and may need modification to meet an individual awarding body's requirements and locally available resources. The first 10 assignments satisfy the requirements for the level 2 courses while the remaining 11 are designed to meet the requirements of level 3 courses. Assignments can normally be carried out in three to five hours, including analysis and report writing and evaluation.

Level 2 assignments

Assignment 1   Electronic circuit construction

For each one of five simple electronic circuits shown in Figs A1.1 to A1.5:

(a) Identify and select the components required to build the circuit.
(b) Identify an appropriate method of construction selected from the following list:

- prototype board;
- tag board;
- strip board;
- printed circuit board;
- wire wrapping.

Assemble and test each circuit according to its circuit diagram. Note that a different construction method must be selected for each circuit.

Assignment 2   Electronic circuit testing

For each of the circuits in Figs A1.1 to A1.5, describe the type and nature of the input and output signals (as appropriate). For each circuit select and use appropriate measuring instruments (e.g. multimeter and oscilloscope) to test each circuit. Write a report to summarize your findings.

Assignment 3   Semiconductor investigation

Prepare a report describing the construction of (a) a junction diode and (b) a bipolar transistor. Describe, in your own words, the principle of operation of each device. With the aid of a simple circuit diagram, describe a typical application for each type of semiconductor.

Figure A1.1   Light operated switch circuit for Assignments 1 and 2

Figure A1.2   Heat-sensing unit circuit for Assignments 1 and 2

Figure A1.3   Power supply circuit for Assignments 1 and 2

Figure A1.4   Astable multivibrator circuit for Assignments 1 and 2

Figure A1.5   Half-wave rectifier circuit for Assignments 1 and 2

Assignment 4   Basic logic functions

Write a report identifying the types and symbols (both BS3939 and MIL/ANSI) used for all basic logic gates (AND, OR, NOT, NAND and NOR). Include in your report a description of the operation of each logic gate together with a truth table.

Assignment 5   Applications of logic circuits

With the aid of labelled diagrams, describe TWO applications of logic gates. One application should be based on combinational logic while the other should use sequential logic.

Assignment 6   Electronic measuring instruments

Write a report describing the operation and use of (a) a multimeter and (b) an oscilloscope. Illustrate your report with records of measurements carried out on three common electronic components and two simple electronic circuits.

Assignment 7   Microprocessors

Investigate TWO common types of microprocessor. Produce an A4-size data sheet for each device giving a brief specification of the chip, package and pin-connecting details, clock frequencies, details of the internal registers, descriptions of bus signals (including the number of address and address lines), etc.

Assignment 8   Monostable timer

Design, construct and test a variable timer based on a 555 device connected in monostable mode. The timer is to be adjustable over the range 10 s to 90 s and is to produce an output for driving a low-voltage relay.

Assignment 9   Square wave generator

Design, construct and test a 5 V 1 kHz square wave generator based on a 555 timer connected in astable mode.

Assignment 10   A dipole aerial

Design, construct and test a half-wave dipole aerial for UHF band IV/V television reception. Select a suitable feeder to connect the aerial to a TV receiver and investigate the directional properties of the aerial. Compare these with what you would expect from radio theory.

Level 3 assignments

Assignment 11   Power supply investigation

With the aid of an electrical specification and operating manual for a typical low-voltage d.c. power supply, write a report explaining the characteristics of the unit. Also explain the meaning of each of the unit's specifications. Carry out a simple load test on the supply, plot a graph to illustrate your results and comment on your findings.

Assignment 12   Amplifier circuit investigation

Write a report that describes and explains one small-signal Class-A discrete amplifier circuit one Class-B power amplifier circuit, and one amplifier circuit based on an operational amplifier. The report should identify and give typical specifications for each type of amplifier. Carry out a simple gain and frequency response test on one of the amplifier circuits, plot a graph to illustrate your results and comment on your findings.

Assignment 13   Small-signal amplifier design and construction

Design, construct and test a single-stage transistor amplifier. Write a report describing the model used for the small-signal amplifier and include a detailed comparison of the predicted characteristics compared with the measured performance of the stage.

Assignment 14   Electronic counter investigation

Prepare a report describing the characteristics of J-K bistable elements when compared with D-type and R-S bistables. Using the data obtained, design, construct and test a four-stage binary counter. Modify the design using a standard logic gate to produce a decade counter. Include in your report timing diagrams for both the binary and decade counters.

Assignment 15   Waveform generator investigation

Prepare a report describing one oscillator circuit, one bistable circuit and one monostable circuit. The report should include a description of one industrial application for each of the circuits together with sample calculations of frequency and pulse rate for the two oscillator circuits, and pulse width for the monostable circuit.

Assignment 16   Performance testing

Carry out performance tests on two different amplifier circuits, two waveform generators and two digital circuits. Compare the measured performance of each circuit with the manufacturer's specification and present your findings in a written report. The report should include details of the calibration and operation of each test instrument in accordance with the manufacturer's handbooks as well as evidence of the adoption of safe working practice.

Assignment 17   Microprocessor clock

Design, construct and test (using an oscilloscope) a microprocessor clock based on two Schmitt inverting logic gates and a quartz crystal. The clock is to produce a square wave output at a frequency of between 1 MHz and 6 MHz (depending upon the quartz crystal used).

Assignment 18   Assembly language programming

Use a simple 8-bit microprocessor system to develop a simple assembly language program that will read the state of a set of eight input switches connected to an input port and illuminate a bank of eight light emitting diodes connected to an output port. The state of the input switches is to be indicated by the light emitting diodes.

Assignment 19   Variable pulse generator

Design, construct and test a variable pulse generator (along the lines of that shown on page 199). The pulse generator is to provide an output from 1 Hz to 10 kHz with a pulse width variable from 50 µs to 500 ms.

Assignment 20   AM radio tuner

Design, construct and test a simple AM radio tuner based on a diode demodulator and a variable tuned circuit. The radio tuner is to cover the medium wave band from 550 kHz to 1.5 MHz and its output is to be connected to an external audio amplifier. Investigate the performance of the tuner and suggest ways in which it could be improved.

Assignment 21   Analogue multimeter

Design, construct and test a simple analogue multimeter based on a 1 mA moving coil meter. The multimeter is to have three voltage ranges (10 V, 50 V and 100 V), two current ranges (10 mA and 50 mA) and one ohms range.

Appendix 2

Revision problems

These 70 problems provide you with a means of checking your understanding prior to an end-of-course assessment or formal examination. If you have difficulty with any of the questions you should refer to the page numbers indicated.

1. A $120\,k\Omega$ resistor is connected to a 6 V battery. Determine the current flowing. [Page 6]
2. A current of 45 mA flows in a resistor of $2.7\,k\Omega$. Determine the voltage dropped across the resistor. [Page 6]
3. A 24 V d.c. supply delivers a current of 1.5 A. Determine the power supplied. [Page 8]
4. A $27\,\Omega$ resistor is rated at 3 W. Determine the maximum current that can safely be applied to the resistor. [Page 8]
5. A load resistor is required to dissipate a power of 50 W from a 12 V supply. Determine the value of resistance required. [Page 8]
6. An electrical conductor has a resistance of $0.05\,\Omega$ per metre. Determine the power wasted in a 175 m length of this conductor when a current of 8 A is flowing in it. [Page 7]
7. Figure A2.1 shows a node in a circuit. Determine the value of I_X. [Page 46]
8. Figure A2.2 shows part of a circuit. Determine the value of V_X. [Page 46]
9. A capacitor of $200\,\mu F$ is charged to a potential of 50 V. Determine the amount of charge stored. [Page 30]
10. A sinusoidal a.c. supply has a frequency of 400 Hz and an r.m.s. value of 120 V. Determine the periodic time and peak value of the supply. [Page 66]
11. Four complete cycles of a waveform occur in a time interval of 20 ms. Determine the frequency of the waveform. [Page 66]
12. Determine the periodic time, frequency and amplitude of each of the waveforms shown in Fig. A2.3. [Page 68]
13. Determine the effective resistance of each circuit shown in Fig. A2.4. [Page 25]
14. Determine the effective capacitance of each circuit shown in Fig. A2.5. [Page 35]
15. Determine the effective inductance of each circuit shown in Fig. A2.6. [Page 40]
16. A quantity of 100 nF capacitors is available, each rated at 100 V working. Determine how several of these capacitors can be connected to produce an equivalent capacitance of: (a) 50 nF rated at 200 V; (b) 250 nF rated at 100 V; and (c) 300 nF rated at 100 V. [Page 35]
17. Two 60 mH inductors and two 5 mH inductors are available, each rated at 1 A. Determine how some or all of these can be connected to produce an equivalent inductance of: (a) 30 mH rated at 2 A; (b) 40 mH rated at 1 A; and (c) 125 mH rated at 1 A. [Page 40]
18. Determine the resistance looking into the network shown in Fig. A2.7, (a) with C and D open-circuit and (b) with C and D shorted together. [Page 25]

Figure A2.1

Figure A2.2

Figure A2.3

Figure A2.4

19. Determine the current flowing in each resistor and voltage dropped across each resistor in Fig. A2.8. [Page 46]
20. Determine the current flowing in the voltmeter movement shown in Fig. A2.9. [Page 50]
21. Assuming that the capacitor shown in Fig. A2.10 is initially fully discharged (by switching to position B), determine the current in $R1$ at the instant that S1 is switched to position A.

Also determine the capacitor voltage 1 minute after operating the switch. [Page 53]
22. Determine the time taken for the output voltage in Fig. A2.11 to reach 4 V after the arrival of the pulse shown (assume that the capacitor is initially uncharged). [Page 53]
23. In Fig. A2.12 determine the current supplied to the inductor 100 ms after pressing the 'start' button. [Page 59]
24. Determine the reactance at 2 kHz of (a) a 60 mH inductor and (b) a 47 nF capacitor. [Page 69]
25. A 50 µF capacitor is connected to a 12V, 50 Hz a.c. supply. Determine the current flowing. [Page 69]
26. An inductor of 2 H is connected to a 12V, 50 Hz a.c. supply. If the inductor has a winding resistance of 40 Ω, determine the current flowing and the phase angle between the supply voltage and supply current. [Page 71]

Figure A2.5

Figure A2.6

Figure A2.7

Figure A2.8

Figure A2.9

Figure A2.10

Figure A2.11

Figure A2.12

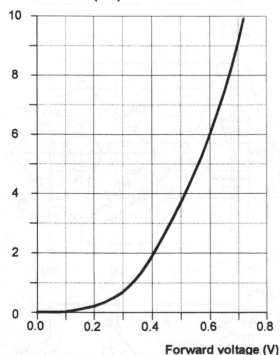

Figure A2.13

27. An inductor of 100 μH is connected in series with a variable capacitor. If the capacitor is variable over the range 50 pF to 500 pF,

determine the maximum and minimum values of resonant frequency for the circuit. [Page 74]

28. An audio amplifier delivers an output power of 40 W r.m.s. to an 8 Ω resistive load. What r.m.s. voltage will appear across the load? [Page 8]

29. A transformer has 400 primary turns and 60 secondary turns. The primary is connected to a 220 V a.c. supply and the secondary is connected to a load resistance of 20 Ω. Assuming that the transformer is perfect, determine: (a) the secondary voltage; (b) the secondary current; and (c) the primary current. [Page 75]

30. Figure A2.13 shows the characteristic of a diode. Determine the resistance of the diode when (a) $V_F = 2\,V$ and (b) $I_F = 9\,mA$. [Page 84]

31. A transistor operates with a collector current of 25 mA and a base current of 200 μA. Determine: (a) the value of emitter current; (b) the value of common-emitter current gain; and (c) the new value of collector current if the base current increases by 50%. [Page 93]

32. A zener diode rated at 5.6 V is connected to a 12 V d.c. supply via a fixed series resistor of 56 Ω. Determine the current flowing in the resistor, the power dissipated in the resistor and the power dissipated in the zener diode. [Page 112]

33. An amplifier has identical input and output resistances and provides a voltage gain of 26 dB. Determine the output voltage produced if an input of 50 mV is applied. [Page 265]

34. Figure A2.14 shows the frequency response of an amplifier. Determine the mid-band voltage gain and the upper and lower cut-off frequencies. [Page 119]

35. Figure A2.15 shows the frequency response of an amplifier. Determine the bandwidth of the amplifier. [Page 120]

36. The transfer characteristic of a transistor is shown in Fig. A2.16. Determine (a) the static value of common-emitter current gain at $I_C = 50\,mA$ and (b) the dynamic (small-signal) value of common-emitter current gain at $I_C = 50\,mA$. [Page 127]

37. The output characteristics of a bipolar transistor are shown in Fig. A2.17. If the transistor operates with $V_{CC} = 15\,V$, $R_L = 500\,\Omega$ and $I_B = 40\,\mu A$ determine:

Figure A2.14

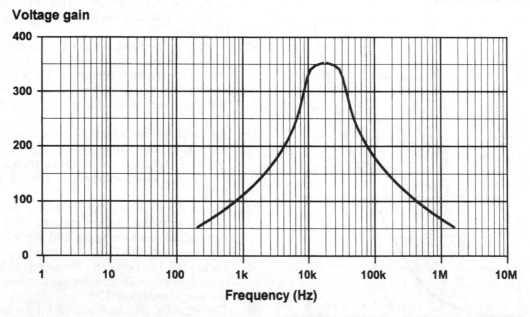

Figure A2.15

(a) the quiescent value of collector–emitter voltage;
(b) the quiescent value of collector current;
(c) the peak–peak output voltage produced by a base input current of 40 μA. [Page 131]

38. The output characteristics of a field effect transistor are shown in Fig. A2.18. If the transistor operates with $V_{DD} = 18\,V$, $R_L = 3\,k\Omega$ and $V_{GS} = -1.5\,V$ determine:

(a) the quiescent value of drain–source voltage;

Figure A2.16

Figure A2.17

(b) the quiescent value of drain current;
(c) the peak–peak output voltage produced by a gate input voltage of 1 V pk–pk;
(d) the voltage gain of the stage. [Page 130]

39. Figure A2.19 shows the circuit of a common-emitter amplifier stage. Determine the values of I_B, I_C, I_E and the voltage at the emitter. [Page 130]

Figure A2.18

40. A transistor having $h_{ie} = 2.5\,k\Omega$ and $h_{fe} = 220$ is used in a common-emitter amplifier stage with $R_L = 3.3\,k\Omega$. Assuming that h_{oe} and h_{re} are negligible, determine the voltage gain of the stage. [Page 128]

41. An astable multivibrator is based on coupling capacitors $C1 = C2 = 10\,nF$ and timing resistors $R1 = 10\,k\Omega$ and $R2 = 4\,k\Omega$.

Figure A2.19

Determine the frequency of the output signal. [Page 155]

42. A sine wave oscillator is based on a Wien bridge with $R = 5\,k\Omega$ and $C = 15\,nF$. Determine the frequency of the output signal. [Page 153]

43. The frequency response characteristic of an operational amplifier is shown in Fig. A2.20. If the device is configured for a closed-loop gain of 200, determine the resulting bandwidth. [Page 142]

44. Redraw Fig. A2.21 using American (MIL/ANSI) symbols. [Page 164]

45. Draw the truth table for the logic gate arrangement shown in Fig. A2.22. [Page 165]

46. Redraw Fig. A2.23 using BS symbols. [Page 164]

47. What single logic gate can be used to replace the logic circuit shown in Fig. A2.24? [Page 165]

48. What single logic gate can be used to replace the logic circuit shown in Fig. A2.25? [Page 165]

49. Devise arrangements of logic gates that will produce the truth tables shown in Fig. A2.26. Use the minimum number of logic gates in each case. [Page 165]

Figure A2.20

Figure A2.21

Figure A2.23

Figure A2.24

Figure A2.22

Figure A2.25

A	B	Y
0	0	1
0	1	1
1	0	0
1	1	1

A	B	C	Y
0	0	0	0
0	0	1	0
0	1	0	0
0	1	1	1
1	0	0	0
1	0	1	0
1	1	0	1
1	1	1	1

A	B	C	Y
0	0	0	1
0	0	1	1
0	1	0	0
0	1	1	1
1	0	0	0
1	0	1	1
1	1	0	0
1	1	1	1

Figure A2.26

Figure A2.27

50. A 1 kHz square wave clock waveform is applied to the circuit shown in Fig. A2.27. Sketch the output waveform against a labelled time axis. [Page 166]
51. (a) Convert 7B hexadecimal to binary.
    (b) Convert 11000011 binary to hexadecimal. [Page 179]
52. What is the largest value, expressed (a) in decimal, and (b) in binary, that can appear at any one time on a 16-bit data bus. [Page 179]
53. Sketch a diagram showing the basic arrangement of a microprocessor system. Label your drawing clearly. [Page 178]
54. (a) Explain the function of a microprocessor clock.
    (b) Explain why a quartz crystal is used to determine the frequency of a microprocessor clock. [Page 184]
55. Sketch the circuit diagram of a typical microprocessor clock. Label your drawing clearly. [Page 183]
56. Explain, briefly, how a microprocessor fetches and executes instructions. Illustrate your answer with a timing diagram showing at least one fetch–execute cycle. [Page 184]
57. Sketch the circuit diagram of a monostable timer based on a 555 device. Explain, briefly, how the circuit operates. [Page 194]
58. A 555 timer is connected in monostable mode. If the values of the timing components used are $C = 100\,\text{nF}$ and $R = 10\,\text{k}\Omega$, determine the monostable pulse time. [Page 194]
59. Determine the frequency of a radio signal that has a wavelength of 1500 m. [Page 202]
60. Determine the wavelength of a radio signal that has a frequency of 40 MHz. [Page 202]

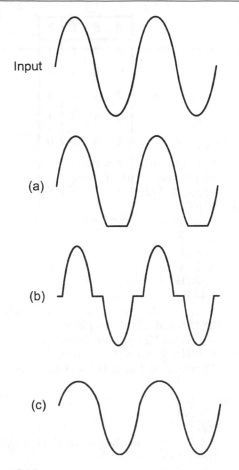

Input

(a)

(b)

(c)

Figure A2.28

61. A superhet medium wave broadcast receiver with an intermediate frequency of 470 kHz is to cover the frequency range 560 kHz to 1.58 MHz. Over what frequency range should the local oscillator be tuned? [Page 207]

62. Explain, with the aid of waveforms, the operation of a simple AM demodulator. [Page 209]

63. Explain, with the aid of a labelled sketch, how the voltage and current are distributed in a half-wave dipole aerial. [Page 210]

64. Determine the length of a half-wave dipole for use at a frequency of 70 MHz. [Page 210]

65. Sketch the block schematic of a simple TRF radio receiver. Briefly explain the function of each stage. [Page 206]

66. A moving coil meter has a full-scale deflection current of 1 mA and a coil resistance of 400 Ω. Determine the value of the multiplier resistor if the meter is to be used as a voltmeter reading 0 to 15 V. [Page 216]

67. A moving coil meter has a full-scale deflection current of 5 mA and a coil resistance of 120 Ω. Determine the value of shunt resistor if the meter is to be used as an ammeter reading 0 to 20 mA. [Page 217]

68. Explain the term 'ohms-per-volt' as applied to a voltmeter. [Page 219]

69. Sketch the circuit of a simple ohmmeter based on a moving coil meter. [Page 217]

70. Identify each of the forms of distortion shown in Fig. A2.28. [Page 240]

Appendix 3

Answers to problems

Chapter 1
1.1   coloumbs, joules, hertz
1.2   3.6 MJ
1.3   0.52 radian
1.4   11.46°
1.5   39.57 kΩ
1.6   0.68 H
1.7   2.45 nF
1.8   0.19 mA
1.9 4.75×10^{-4} V
1.10 16.5×10^{6} Ω
1.11 $4.8 \times 10^{6}, 7.2 \times 10^{3}, 4 \times 10^{3}, 0.5 \times 10^{-3}$
1.12   silver
1.13   33.3 mA
1.14   6.72 V
1.15   3.3 kΩ
1.16   15 Ω
1.17   0.436 Ω
1.18   0.029 W
1.19   0.675 W
1.20   57.7 mA
1.21 0.625×106 V/m
1.22   12 A
1.23   6 μWb

Chapter 2
2.1   60 Ω, 3.75 W, wirewound
2.2   270 kΩ 5%, 10 Ω 10%, 6.8 MΩ 5%,
       0.39 Ω 5%, 2.2 kΩ 2%
2.3   44.65 Ω to 49.35 Ω
2.4   27 Ω and 33 Ω in series,
       27 Ω and 33 Ω in parallel,
       56 Ω and 68 Ω in series,
       27 Ω 33 Ω and 56 Ω in parallel,
       27 Ω 33 Ω and 68 Ω in series
2.5   66.67 Ω
2.6   10 Ω
2.7   102 Ω, 78.5 Ω
2.8   407.2 Ω
2.9   98.7 kΩ
2.10 3.21×10^{-4}
2.11   3.3 μF and 4.7 μF in parallel,
       1 μF and 10 μF in parallel,
       1 μF 3.3 μF 4.7 μF and 10 μF in parallel,

1 μF and 10 μF in series,
3.3 μF and 4.7 μF in series
2.12   60 pF, 360 pF
2.13   50 pF
2.14   20.79 mC
2.15   1.98 nF
2.16   69.4 μF
2.17   0.313 V
2.18   0.136 H
2.19   0.48 J
2.20   10 mH 22 mH and 60 mH in parallel,
       10 mH and 22 mH in parallel,
       10 mH and 22 mH in series,
       10 mH and 60 mH in series,
       10 mH 60 mH and 100 mH in series

Chapter 3
3.1   275 mA
3.2   200 Ω
3.3   1.5 A away from the junction, 215 mA away
       from the junction
3.4   1.856 V, 6.6 V
3.5   0.1884 A, 0.1608 A, 0.0276 A, 5.09 V,
       0.91 V, 2.41 V
3.6   1.8 V, 10.2 V
3.7   0.5 A, 1.5 A
3.8   1 V, 2 V, 3 V, 4 V, 5 V
3.9   1.5 kΩ, 60 mA
3.10   40.6 ms
3.11   3.54 s
3.12   112.1 μF
3.13   0.128 s
3.14   2.625 V, 5 Ω
3.15   21 V, 7 Ω, 3 A, 7 Ω
3.16   50 mA, 10 V
3.17   50 V, 10 V

Chapter 4
4.1   4 ms, 35.35 V
4.2   59.88 Hz, 3394 V
4.3   50 Hz 30 V pk–pk, 15 Hz 10 V pk–pk,
       150 kHz 0.1 V pk–pk
4.4   19 V, −11.8 V
4.5   10.6 V

4.6   36.19 kΩ, 144.76 Ω
4.7   3.54 mA
4.8   10.362 Ω, 1.45 kΩ
4.9   4.71 V
4.10   592.5 Ω, 0.186 A
4.11   0.55, 0.487 A
4.12   157 nF
4.13   1.77 MHz to 7.58 MHz
4.14   7.5 mA, 2.71 V
4.15   281 kHz, 41.5, 6.77 kHz
4.16   18 V
4.17   245 V

Chapter 5
5.1   silicon (forward threshold appx. 0.6 V)
5.2   41 Ω, 150 Ω
5.4   9.1 V zener diode
5.5   250 Ω
5.6   germanium low-power high-frequency,
       silicon low-power low-frequency,
       silicon high-power low-frequency,
       silicon low-power high frequency
5.7   2.625 A, 20
5.8   5 mA, 19.6
5.9   16.7
5.10   BC108
5.11   8 μA, 1.1 mA
5.12   47 mA, 94, 75
5.13   12.5 mA, 12 V, 60 μA
5.14   16 mA

Chapter 6
6.1   80 mV
6.2   5 mV
6.3   200 Ω
6.4   12.74 V, 9.1 V, 8.4 V
6.5   36.4 mA, 0.33 W
6.6   0 V, 12.04 V, 0 V
6.7   0.5 Ω, 8.3 V
6.8   1%, 15.15 V

Chapter 7
7.1   40, 160, 6400, 100 Ω
7.2   2 V
7.3   56, 560 kHz, 15 Hz
7.4   18.5
7.5   0.0144
7.6   2.25 V
7.7   13 μA, 3.39 V, 2.7 V, 4.51 V
7.8   5 V, 7 mA, 8.5 V
7.9   12.2 V, 6.1 mA, 5.5 V

Chapter 8
8.1   10 V
8.2   40 dB, 600 kHz
8.3   200 kΩ
8.4   +1 V, −1 V, 0 V, −2 V, +2 V, 0 V
8.5   4 kHz, 100 Hz
8.6   11, 1
8.7   10, 3.38 kHz, 338 kHz

Chapter 9
9.1   4.44, 40
9.2   6.49 kΩ
9.3   18 kΩ
9.4   5.63 V pk–pk
9.5   14.3 kΩ, 42.9 kΩ

Chapter 10
10.9   Low-power Schottky (LS) TTL,
       27th Month of 1989
10.10 0.6 V

Chapter 11
11.1   00111010
11.2   C2
11.3   (c)
11.4   1 048 576
11.5   1024
11.6   +127
11.7   (a) Input port, FEH
            Output port, FFH
       (b) 001010000

Chapter 12
12.5   (a) *VR2*
       (b) *S2*
       (c) *R4*
       (d) *VR1*
       (e) *S1*
       (f) *R2*
       (g) *VR3*
       (h) *R5*
       (i) *C10*
       (j) *C5* and *R3*
12.6   50 Ω

Chapter 13
13.1   1.58 m
13.2   23.1 MHz
13.3   4 m
13.4   1.02 MHz to 2.07 MHz
13.5   10.7 MHz, 170 kHz, 63
13.6   (a) *L1/C2*
       (b) *R1/R2*

      (c) *L2*

      (d) *R4/C5*

      (e) *VR1*

      (f) *C6*

      (g) *C7*

      (h) *C1*

13.7   3 m

13.8   24 W

Chapter 14

14.1   19.6 kΩ

14.2   11.11 Ω

14.3   16.67 kΩ

14.4   100 kΩ

14.5   (a) 19.99 V

      (b) 10 mV

14.6   25 kΩ

14.7   60 mA

14.8   35.3 µA

14.9   (a) 3.33 ms, 4 V

      (b) 125 ns, 150 mV

14.10 (a) 7.8 µs

      (b) 3.4 µs

      (c) 4.4 µs

      (d) 1.5 µs

      (e) 1 µs

      (f) 5 V

Appendix 4

Semiconductor pin connections

Signal and rectifier diodes

1N914
OA90
OA91

1N4148

Yellow   Brown   Yellow   Grey

1N4001
etc

Power rectifier diodes

1N5401
etc

Zener diodes

BZY88

BZX61
BZX85

Small-signal transistors

BC107
BC108
BC109

2N3819
BF244

BC547
BC548
BC549

BC184L
BC212L

Medium power and driver transistors

BFY50
BFY51
BFY52
2N3053

Power transistors

TIP31

2N3053

BD131
BD132

Figure A4.1   Diodes

Figure A4.2   Transistors

Figure A4.3   Integrated circuits

Figure A4.4

Appendix 5

Decibels

Decibels (dB) are a convenient means of expressing gain (amplification) and loss (attenuation) in electronic circuits. In this respect, they are used as a **relative** measure (i.e. comparing one voltage with another, one current with another, or one power with another). In conjunction with other units, decibels are sometimes also used as an **absolute** measure. Hence dBV are decibels relative to $1\,V$, dBm are decibels relative to $1\,mW$, etc.

The decibel is one-tenth of a bel which, in turn, is defined as the logarithm, to the base 10, of the ratio of output power (P_{out}) to input power (P_{in}).

Gain and loss may be expressed in terms of power, voltage and current such that:

$$A_P = \frac{P_{out}}{P_{in}} \quad A_V = \frac{V_{out}}{V_{in}} \quad \text{and} \quad A_I = \frac{I_{out}}{I_{in}}$$

where A_P, A_V or A_I is the power, voltage or current gain (or loss) expressed as a ratio, P_{in} and P_{out} are the input and output powers, V_{in} and V_{out} are the input and output voltages, and I_{in} and I_{out} are the input and output currents. Note, however, that the powers, voltages or currents should be expressed in the same units/multiples (e.g. P_{in} and P_{out} should both be expressed in W, mW, μW or nW).

It is often more convenient to express gain in decibels (rather than as a simple ratio) using the following relationships:

$$A_P = 10\log_{10}\frac{(P_{out})}{(P_{in})} \quad A_V = 20\log_{10}\frac{(V_{out})}{(V_{in})}$$

$$\text{and} \quad A_I = 20\log_{10}\frac{(I_{out})}{(I_{in})}$$

Note that a positive result will be obtained whenever P_{out}, V_{out}, or I_{out} is greater than P_{in}, V_{in}, or I_{in}, respectively. A negative result will be obtained whenever P_{out}, V_{out}, or I_{out} is less than P_{in}, V_{in} or I_{in}. A negative result denotes attenuation rather than amplification. A negative gain is thus equivalent to an attenuation (or loss). If desired, the formulae may be adapted to produce a positive result for attenuation simply by inverting the ratios, as shown below:

$$A_P = 10\log_{10}\frac{(P_{in})}{(P_{out})} \quad A_V = 20\log_{10}\frac{(V_{in})}{(V_{out})}$$

$$\text{and} \quad A_I = 20\log_{10}\frac{(I_{in})}{(I_{out})}$$

where A_P, A_V or A_I is the power, voltage or current gain (or loss) expressed in decibels, P_{in} and P_{out} are the input and output powers, V_{in} and V_{out} are the input and output voltages, and I_{in} and I_{out} are the input and output currents. Note, again, that the powers, voltages or currents should be expressed in the same units/multiples (e.g. P_{in} and P_{out} should both be expressed in W, mW, μW or nW).

It is worth noting that, for identical decibel values, the values of voltage and current gain can be found by taking the square root of the corresponding value of power gain. As an example, a voltage gain of $20\,dB$ results from a voltage ratio of 10 while a power gain of $20\,dB$ corresponds to a power ratio of 100.

Finally, it is essential to note that the formulae for voltage and current gain are only meaningful when the input and output impedances (or resistances) are identical. Voltage and current gains expressed in decibels are thus only valid for matched (constant impedance) systems.

The following table gives some useful decibel values:

Decibels (dB)	Power gain (ratio)	Voltage gain (ratio)	Current gain (ratio)
0	1	1	1
1	1.26	1.12	1.12
2	1.58	1.26	1.26
3	2	1.41	1.41
4	2.51	1.58	1.58
5	3.16	1.78	1.78
6	3.98	2	2
7	5.01	2.24	2.24
8	6.31	2.51	2.51
9	7.94	2.82	2.82
10	10	3.16	3.16
13	19.95	3.98	3.98
16	39.81	6.31	6.31

Decibels (dB)	Power gain (ratio)	Voltage gain (ratio)	Current gain (ratio)
20	100	10	10
30	1000	31.62	31.62
40	10 000	100	100
50	100 000	316.23	316.23
60	1 000 000	1000	1000
70	10 000 000	3162.3	3162.3

Example A5.1

An amplifier with matched input and output resistances provides an output voltage of 1 V for an input of 25 mV. Express the voltage gain of the amplifier in decibels.

Solution

The voltage gain can be determined from the formula:

$$A_V = 20 \log_{10}(V_{out}/V_{in})$$

where $V_{in} = 25$ mV and $V_{out} = 1$ V.
  Thus:

$$A_V = 20 \log_{10}(1\,V/25\,mV) = 20 \log_{10}(40)$$
$$= 20 \times 1.6 = 32\,dB$$

Example A5.2

A matched $600\,\Omega$ attenuator produces an output of 1 mV when an input of 20 mV is applied. Determine the attenuation in decibels.

Solution

The attenuation can be determined by applying the formula:

$$A_V = 20 \log_{10}(V_{in}/V_{out})$$

where $V_{in} = 20$ mV and $V_{out} = 1$ mV.
  Thus:

$$A_V = 20 \log_{10}(20\,mV/1\,mV) = 20 \log_{10}(20)$$
$$= 20 \times 1.3 = 26\,dB$$

Example A5.3

An amplifier provides a power gain of 33 dB. What output power will be produced if an input of 2 mW is applied?

Solution

Here we must re-arrange the formula to make P_{out} the subject, as follows:

$$A_P = 10 \log_{10}(P_{out}/P_{in})$$

thus

$$A_P/10 = \log_{10}(P_{out}/P_{in})$$

or

$$\text{antilog}_{10}(A_P/10) = P_{out}/P_{in}$$

Hence

$$P_{out} = P_{in} \times \text{antilog}_{10}(A_P/10)$$

Now $P_{in} = 2$ mW $= 20 \times 10^{-3}$ W and $A_P = 33$ dB, thus

$$P_{out} = 2 \times 10^{-3} \times \text{antilog}_{10}(33/10)$$
$$= 2 \times 10^{-3} \times \text{antilog}_{10}(3.3)$$
$$= 2 \times 10^{-3} \times 1.995 \times 10^{-3} = 3.99\,W$$

Appendix 6

Mathematics for electronics

This section introduces the mathematical techniques that are essential to developing a good understanding of electronics. The content is divided into seven sections: notation, algebra, equations, graphs, trigonometry, Boolean algebra, logarithms and exponential growth and decay. We have included a number of examples and typical calculations.

Notation

The standard notation used in mathematics provides us with a shorthand that helps to simplify the writing of mathematical expressions. Notation is based on the use of symbols that you will already recognize. These include $=$ (equals), $+$ (addition), $-$ (subtraction), $\times$ (multiplication), and $\div$ (division). Other symbols that you may not be so familiar with include $<$ (less than), $>$ (greater than), $\propto$ (proportional to), and $\sqrt{}$ (square root).

Indices

The number 4 is the same as 2×2, that is, 2 multiplied by itself. We can write (2×2) as 2^2. In words, we would call this 'two raised to the power two' or simply 'two squared'. Thus:

$$2 \times 2 = 2^2$$

By similar reasoning we can say that:

$$2 \times 2 \times 2 = 2^3$$

and

$$2 \times 2 \times 2 \times 2 = 2^4$$

In these examples, the number that we have used (i.e. 2) is known as the *base* whilst the number that we have raised it to is known as an *index*. Thus, 2^4 is called 'two to the power of four', and it consists of a base of 2 and an index of 4. Similarly, 5^3 is called 'five to the power of three' and has a base of 5 and an index of 3. Special names are used when the indices are 2 and 3, these being called 'squared'

and 'cubed', respectively. Thus 7^2 is called 'seven squared' and 9^3 is called 'nine cubed'. When no index is shown, the power is 1, i.e. 2^1 means 2.

Reciprocals

The *reciprocal* of a number is when the index is -1 and its value is given by 1 divided by the base. Thus the reciprocal of 2 is 2^{-1} and its value is 1/2 or 0.5. Similarly, the reciprocal of 4 is 4^{-1} which means 1/4 or 0.25.

Square roots

The *square root* of a number is when the index is 1/2. The square root of 2 is written as $2^{1/2}$ or $\sqrt{2}$. The value of a square root is the value of the base which when multiplied by itself gives the number. Since $3 \times 3 = 9$, then $\sqrt{9} = 3$. However, $(-3) \times (-3) = 9$, so we have a second possibility, i.e. $\sqrt{9} = \pm 3$. There are always two answers when finding the square root of a number and we can indicate this is by placing a $\pm$ sign in front of the result meaning 'plus or minus'. Thus:

$$4^{1/2} = \sqrt{4} = \pm 2$$

and

$$9^{1/2} = \sqrt{9} = \pm 3$$

Variables and constants

Unfortunately, we don't always know the value of a particular quantity that we need to use in a calculation. In some cases the value might actually change, in which case we refer to it as a *variable*. In other cases, the value might be fixed but we might prefer not to actually quote its value. In this case we refer to the value as a *constant*.

An example of a *variable quantity* is the output voltage produced by a power supply when the load current is changing. An example of a *constant quantity* might be the speed at which radio waves travel in space.

In either case, we use a symbol to represent the quantity. The symbol itself (often a single letter) is a form of shorthand notation. For example, in the case of the voltage from the power supply we would probably use v to represent *voltage*. Whereas, in the case of the speed of travel of a radio wave we normally use c to represent the speed of radio waves in space (i.e. the speed of light).

We can use letters to represent both variable and constant quantities in mathematical notation. For example, the statement:

voltage is equal to *current* multiplied by *resistance*

can be written in mathematical notation as follows:

$$V = I \times R$$

where V represents voltage, I represents current, and R represents resistance.

Similarly, the statement:

resistance is equal to *voltage* divided by *current*

can be written in mathematical notation as follows:

$$R = \frac{V}{I}$$

where R represents resistance, V represents voltage, and I represents current.

Proportionality

In electronic circuits, when one quantity changes it normally affects a number of other quantities. For example, if the output voltage produced by a power supply increases when the resistance of the load to which it is connected remains constant, the current supplied to the load will also increase. If the output voltage doubles, the current will also double in value. If, on the other hand, the output voltage falls by 50% the current will also fall by 50%. To put this in a mathematical way we can say that (provided that load resistance remains constant):

load current is *directly proportional* to output voltage

Using mathematical notation and symbols to represent the quantities, we would write this as follows:

$$I \propto V$$

where I represents load current and V represents output voltage.

In some cases, an increase in one quantity will produce a *reduction* in another quantity. For example, if the frequency of a wavelength of a radio wave increases its wavelength is reduced. To put this in a mathematical way we say that (provided the speed of travel remains constant):

wavelength is *inversely proportional* to frequency

Using mathematical notation and symbols to represent the quantities, we would write this as follows:

$$T \propto \frac{1}{f}$$

where T represents periodic time and f represents frequency.

It's useful to illustrate these important concepts using some examples.

Example 1

The power in a loudspeaker is proportional to the square of the r.m.s. output voltage and inversely proportional to the impedance of the speaker. Write down an equation for the power, P, in terms of the voltage, V, and impedance, Z. Obtain a formula for P in terms of V and Z.

From the information given, we can say that:

$$P \propto V^2 \text{ (power is proportional to voltage squared)} \tag{1}$$

and

$$P \propto \frac{1}{Z} \text{ (power is inversely proportional to impedance)} \tag{2}$$

We can rewrite this expression using a negative index, as follows:

$$P \propto Z^{-1}$$

We can go one stage further and combine these two relationships to obtain a formula for the power:

$$P = V^2 Z^{-1} = \frac{V^2}{Z}$$

Example 2

In Example 1, determine the power delivered to the loudspeaker if the r.m.s. voltage. V, is 8 V and the loudspeaker has an impedance, Z, of 4 Ω.

$$P = V^2 Z^{-1} = \frac{V^2}{Z} = \frac{8^2}{4} = \frac{8 \times 8}{4} = \frac{64}{4} = 16\,\text{W}$$

Example 3

The equipment manufacturer in Example 2 recommends testing the amplifier at a power of 10 W. Determine the r.m.s. voltage at the output that corresponds to this power level.

Here we must rearrange the formula in order to make V the subject. If we multiply both sides of the equation by Z we will have P and Z on one side of the equation and V^2 on the other:

$$P \times Z = \frac{V^2}{Z} \times Z \quad \text{from which}$$

$$PZ = V^2 \frac{Z}{Z} \quad \text{or} \quad PZ = V^2 \tag{1}$$

Rearranging (1) gives : $V^2 = PZ$ (2)

Taking square roots of *both* sides:

$$\sqrt{V^2} = \sqrt{PZ}$$

The square root of V^2 is V hence: $V = \sqrt{PZ} = (PZ)^{1/2}$

We need to find V when $P = 10\,\text{W}$ and $Z = 4\Omega$:

$$V = \sqrt{10 \times 4} = \sqrt{40} = 6.32\,\text{V}$$

Laws of indices

When simplifying calculations involving indices, certain basic rules or laws can be applied, called the *laws of indices*. These are listed below:

(a) When multiplying two or more numbers having the same base, the indices are added. Thus $2^2 \times 2^4 = 2^{2+4} = 2^6$.
(b) When a number is divided by a number having the same base, the indices are subtracted. Thus $2^5/2^2 = 2^{5-2} = 2^3$.
(c) When a number which is raised to a power is raised to a further power, the indices are multiplied. Thus $(2^5)^2 = 2^{5\times 2} = 2^{10}$.
(d) When a number has an index of 0, its value is 1. Thus $2^0 = 1$.

(e) When a number is raised to a negative power, the number is the reciprocal of that number raised to a positive power. Thus $2^{-4} = 1/2^4$. Similarly, $1/2^{-3} = 2^3$.
(f) When a number is raised to a fractional power the denominator of the fraction is the root of the number and the numerator is the power. Thus $4^{3/4} = \sqrt[4]{4^2} = (2)^2 = 4$ and $25^{1/2} = \sqrt{25^1} = \pm 5$.

Standard form

A number written with one digit to the left of the decimal point and multiplied by 10 raised to some power is said to be written in *standard form*. Thus: 1234 is written as 1.234×10^3 in standard form, and 0.0456 is written as 4.56×10^{-2} in standard form.

When a number is written in standard form, the first factor is called the *mantissa* and the second factor is called the *exponent*. Thus the number 6.8×10^3 has a mantissa of 6.8 and an exponent of 10^3.

Numbers having the same exponent can be added or subtracted in standard form by adding or subtracting the mantissae and keeping the exponent the same. Thus:

$$2.3 \times 10^4 + 3.7 \times 10^4 = (2.3 + 3.7) \times 10^4$$
$$= 6.0 \times 10^4,$$

and

$$5.7 \times 10^{-2} - 4.6 \times 10^{-2} = (5.7 - 4.6) \times 10^{-2}$$
$$= 1.1 \times 10^{-2}$$

When adding or subtracting numbers it is quite acceptable to express one of the numbers in non-standard form, so that both numbers have the same exponent. This makes things much easier as the following example shows:

$$2.3 \times 10^4 + 3.7 \times 10^3 = 2.3 \times 10^4 + 0.37 \times 10^4$$
$$= (2.3 + 0.37) \times 10^4$$
$$= 2.67 \times 10^4$$

Alternatively,

$$2.3 \times 10^4 + 3.7 \times 10^3 = 23\,000 + 3700$$
$$= 26\,700 = 2.67 \times 10^4$$

The laws of indices are used when multiplying or dividing numbers given in standard form. For example,

$$(22.5 \times 10^3) \times (5 \times 10^2) = (2.5 \times 5) \times (10^{3+2})$$
$$= 12.5 \times 10^5 \quad \text{or} \quad 1.25 \times 10^6$$

Example 4

Period, t, is the reciprocal of frequency, f. Thus $t = f^{-1} = \dfrac{1}{f}$.

Calculate the period of a radio frequency signal having a frequency of 2.5 MHz.

Now $f = 2.5$ MHz. Expressing this in standard form, $f = 2.5 \times 10^6$ Hz.

$$\text{Since } t = f^{-1} = \frac{1}{2.5 \times 10^6} = \frac{10^{-6}}{2.5} = \frac{1}{2.5} \times 10^{-6}$$
$$= 0.4 \times 10^{-6} = \mathbf{4 \times 10^{-7}\, s}$$

Example 5

Resistors of $3.9\,k\Omega$, $5.6\,k\Omega$ and $10\,k\Omega$ are connected in parallel as shown in Fig. A6.1. Calculate the effective resistance of the circuit.

The resistance of a parallel circuit is given by the equation:

$$\frac{1}{R} = \frac{1}{R1} + \frac{1}{R2} + \frac{1}{R3}$$

Now we know that:

$$R1 = 3.9\,k\Omega = 3.9 \times 10^3\,\Omega$$
$$R2 = 5.6\,k\Omega = 5.6 \times 10^3\,\Omega$$
$$R3 = 10\,k\Omega = 1 \times 10^4\,\Omega = 10 \times 10^3\,\Omega$$

Hence:

$$\frac{1}{R} = \frac{1}{3.9 \times 10^3} + \frac{1}{5.6 \times 10^3} + \frac{1}{10 \times 10^3}$$
$$= \frac{10^{-3}}{3.9} + \frac{10^{-3}}{5.6} + \frac{10^{-3}}{10} = \left(\frac{1}{3.9} + \frac{1}{5.6} + \frac{1}{10}\right) \times 10^{-3}$$
$$= (0.256 + 0.179 + 0.1) \times 10^{-3} = 0.535 \times 10^{-3}$$

Now since $\dfrac{1}{R} = 0.535 \times 10^3$ we can invert both sides of the equation so that:

$$R = \frac{1}{0.535 \times 10^{-3}} = 1.87 \times 10^3 = \mathbf{1.87\,k\Omega}$$

Equations

We frequently need to solve equations in order to find the value of an unknown quantity. Any

Figure A6.1   See Example 5

arithmetic operation may be applied to an equation as long as the equality of the equation is maintained. In other words, the same operation must be applied to both the left-hand side (LHS) and the right-hand side (RHS) of the equation.

Example 6

A current of 0.5 A flows in the $56\,\Omega$ shown in Fig. A6.2. Given that $V = IR$ determine the voltage that appears across the resistor.

It's a good idea to get into the habit of writing down what you know before attempting to solve an equation. In this case:

$$I = 0.5\,A$$
$$R = 56\,\Omega$$
$$V = ?$$

Now

$$V = IR = 0.5 \times 56 = \mathbf{28\,V}$$

Example 7

The current present at a junction in a circuit is shown in Fig. A6.3. Determine the value of the unknown current, I_x.

From Kirchhoff's current law, the algebraic sum of the current at a junction in a circuit is zero. Adopting the convention that current flowing

Figure A6.2   See Example 6

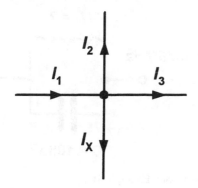

Figure A6.3 See Example 7

towards the junction is positive and that flowing away from the junction is negative we can construct a formula along the following lines:

$$0 = -I_x + I_1 - I_2 - I_3$$

We can rearrange the equation to make I_x the subject by adding I_x to both sides:

$$0 + I_x = -I_x + I_x + I_1 - I_2 - I_3$$

from which:

$$I_x = I_1 - I_2 - I_3$$

Now $I_1 = 0.1\,\text{A}$, $I_2 = 0.25\,\text{A}$ and $I_3 = 0.3\,\text{A}$.

Thus:

$$I_x = 0.1 - 0.25 - 0.3 = -0.45\,\text{A}$$

Hence $I_x = $ **0.45 A flowing towards the junction**.

It is important to note here that our original 'guess' as to the direction of I_x as marked on Fig. A6.3 was incorrect and the current is actually flowing the other way!

Note that it is always a good idea to check the solution to an equation by substituting the solution back into the original equation. The equation should then balance such that the left-hand side (LHS) can be shown to be equal to the right-hand side (RHS).

In many cases it can be more convenient to change the subject of a formula *before* you insert values. The next few examples show how this is done.

Example 8

A copper wire has a length l of 1.5 km, a resistance R of $5\,\Omega$ and a resistivity ρ of $17.2 \times 10^{-6}\,\Omega\,\text{mm}$. Find the cross-sectional area, a, of the wire, given that $R = \dfrac{\rho l}{a}$.

Once again, it is worth getting into the habit of summarizing what you know from the question and what you need to find (don't forget to include the units):

$$R = 5\,\Omega$$

$$\rho = 17.2 \times 10^{-6}\,\Omega\,\text{mm}$$

$$l = 1500 \times 10^3\,\text{mm}$$

$$a = ?$$

Since $R = \dfrac{\rho l}{a}$ then

$$5 = \frac{(17.2 \times 10^{-6})(1500 \times 10^3)}{a}$$

Cross multiplying (i.e. exchanging the '5' for the 'a') gives:

$$a = \frac{(17.2 \times 10^{-6})(1500 \times 10^3)}{5}$$

Now group the numbers and the powers of ten as shown below:

$$a = \frac{17.2 \times 1500 \times 10^{-6} \times 10^3}{5}$$

Next simplify as far as possible:

$$a = \frac{17.2 \times 1500}{5} \times 10^{-6+3}$$

Finally, evaluate the result using your calculator:

$$a = 5160 \times 10^{-3} = 5.16$$

Since we have been working in mm, the result, a, will be in mm^2.

Hence $a = $ **5.16 mm^2**.

It's worth noting from the previous example that we have used the laws of indices to simplify the powers of ten before attempting to use the calculator to determine the final result. The alternative to doing this is to make use of the exponent facility on your calculator. Whichever technique you use it's important to be confident that you are correctly using the exponent notation since it's not unknown for students to produce answers that are incorrect by a factor of 1000 or even 1 000 000 and an undetected error of this magnitude could be totally disastrous!

Before attempting to substitute values into an equation, it's important to be clear about what you know (and what you don't know) and always make sure that you have the correct units. The values marked on components can sometimes be misleading and it's always worth checking that you have interpreted the markings correctly before wasting time solving an equation that doesn't produce the right answer!

Example 9

The reactance, X, of a capacitor is given by $X = \dfrac{1}{2\pi f C}$.

Find the value of capacitance that will exhibit a reactance of $10\,\text{k}\Omega$ at a frequency of $400\,\text{Hz}$.

First, to summarize what we know:

$X = 10\,\text{k}\Omega$

$f = 400\,\text{Hz}$

$\pi = 3.142$ (or use the 'π' button on your calculator!)

$C = ?$

We need to rearrange the formula $X = \dfrac{1}{2\pi f C}$ to make C the subject. This is done as follows:

Cross multiplying gives:

$C = \dfrac{1}{2\pi f \times X}$

(notice how we have effectively 'swapped' the C and the X over).

Next, replacing π, C and f by the values that we know gives:

$C = \dfrac{1}{2 \times 3.142 \times 400 \times 10 \times 10^3} = \dfrac{1}{25\,136 \times 10^3}$

$\quad = \dfrac{1}{2.5136 \times 10^7}$

$\quad = \dfrac{1}{2.5136} \times 10^{-7} = 0.398 \times 10^{-7}\,\text{F}$

Finally, it would be sensible to express the answer in nF (rather than F). To do this, we simply need to multiply the result by 10^9, as follows:

$C = 0.398 \times 10^{-7} \times 10^9 = 0.398 \times 10^{9-7}$

$\quad = 0.398 \times 10^2 = \mathbf{39.8\,nF}$

Example 10

The frequency of resonance, f, of a tuned circuit (see Fig. A6.4) is given by the relationship:

$f = \dfrac{1}{2\pi\sqrt{LC}}$

Figure A6.4   See Example 10

If a tuned circuit is to be resonant at $6.25\,\text{MHz}$ and $C = 100\,\text{pF}$, determine the value of inductance, L.

Here we know that:

$C = 100\,\text{pF} = 100 \times 10^{-12}\,\text{F}$

$f = 6.25\,\text{MHz} = 6.25 \times 10^6\,\text{Hz}$

$\pi = 3.142$ (or use the 'π' button on your calculator!)

$L = ?$

First we will rearrange the formula $f = \dfrac{1}{2\pi\sqrt{LC}}$ in order to make L the subject.

Squaring both sides gives:

$f^2 = \left(\dfrac{1}{2\pi\sqrt{LC}}\right)^2 = \dfrac{1^2}{2^2\pi^2(\sqrt{LC})^2} = \dfrac{1}{4\pi^2 LC}$

Rearranging gives:

$L = \dfrac{1}{4\pi^2 f^2 C}$

We can now replace f, C and π by the values that we know:

$L = \dfrac{1}{4 \times 3.142^2 \times (6.25 \times 10^6)^2 \times 100 \times 10^{-12}}$

$\quad = \dfrac{1}{39.49 \times 39.06 \times 10^{12} \times 100 \times 10^{-12}}$

$\quad = \dfrac{1}{154\,248} = \dfrac{1}{1.54\,248 \times 10^5} = \dfrac{1}{1.54\,248} \times 10^{-5}$

$\quad = 0.648 \times 10^{-5} = 6.48 \times 10^{-6} = \mathbf{6.48\,\mu H}$

More complex equations

More complex equations are solved in essentially the same way as the simple equations that we have just met. Note that equations with square (or higher) laws may have more than one solution. Here is an example.

Example 11

The impedance of the a.c. circuit shown in Fig. A6.5 is given by $Z = \sqrt{R^2 + X^2}$. Determine the reactance, X, of the circuit when $R = 25\,\Omega$ and $Z = 50\,\Omega$.

Here we know that:

$R = 25\,\Omega$

$Z = 50\,\Omega$

$X = ?$

First we will rearrange the formula in order to make X the subject:

$$Z = \sqrt{R^2 + X^2}$$

Squaring both sides gives:

$$Z^2 = \left(\sqrt{R^2 + X^2}\right)^2 = R^2 + X^2$$

Rearranging gives:

$$X^2 = Z^2 - R^2$$

Taking the square root of both sides gives:

$$X = \sqrt{Z^2 - R^2}$$

We can now replace Z and R by the values that we know:

$$X = \sqrt{50^2 - 25^2} = \sqrt{2500 - 625} = \sqrt{1875} = \mathbf{43.3\,\Omega}$$

In the previous example we arrived at a solution of $43.3\,\Omega$ for X. This is not the only solution as we will now show.

As before, we can check that we have arrived at the correct answer by substituting values back into the original equation, as follows:

$R = 25\Omega$

X

$Z = 50\Omega$

Figure A6.5   See Example 11

$$50 = Z = \sqrt{R^2 + X^2} = \sqrt{25^2 + 43.3^2}$$
$$= \sqrt{625 + 1875} = \sqrt{2500} = 50$$

Thus LHS = RHS.

However, we could also have used $-43.3\,\Omega$ for X and produced the *same* result:

$$50 = Z = \sqrt{R^2 + X^2} = \sqrt{25^2 + (-43.3)^2}$$
$$= \sqrt{625 + 1875} = \sqrt{2500} = 50$$

Again, LHS = RHS.

The reason for this apparent anomaly is simply that the result of squaring a negative number will be a positive number (i.e. $(-2)^2 = (-2) \times (-2) = +4$).

Hence, a correct answer to the problem in Example 11 would be:

$X = \mathbf{\pm 43.3\,\Omega}$

Don't panic if the ambiguity of this answer is worrying you! Reactance in an a.c. circuit can either be positive or negative depending upon whether the component in question is an inductor or a capacitor. In this case, the numerical value of the impedance is the same regardless of whether the reactance is *inductive* or *capacitive*.

Graphs

Graphs provide us with a visual way of representing data. They can also be used to show, in a simple pictorial way, how one variable affects another variable. Several different types of graph are used in electronics. We shall start by looking at the most basic type, the straight line graph.

Straight line graphs

Earlier we introduced the idea of *proportionality*. In particular, we showed that the current flowing in a circuit was directly proportional to the voltage applied to it. We expressed this using the following mathematical notation:

$i \propto v$

where i represents load current and v represents output voltage.

We can illustrate this relationship using a simple graph showing current, i, plotted against voltage, v. Let's assume that the voltage applied to the circuit varies over the range $1\,V$ to $6\,V$ and the circuit has a resistance of $3\,\Omega$. By taking a set of measurements of v and i (see Fig. A6.6) we would obtain the following table of corresponding values:

Voltage, v (V)	1	2	3	4	5	6
Current, i (A)	0.33	0.66	1.0	1.33	1.66	2.0

The resulting graph is shown in Fig. A6.7. To obtain the graph, a point is plotted for each pair of corresponding values for v and i. When all the points have been drawn they are connected together by drawing a line. Notice that, in this case, the line that connects the points together takes the form of a straight line. This is *always* the case when two variables are directly proportional to one another.

It is conventional to show the *dependent variable* (in this case it is current, i) plotted on the vertical axis and the *independent variable* (in this case it is voltage, v) plotted on the horizontal axis. If you find these terms a little confusing, just remember that, what you know is usually plotted on the horizontal scale whilst what you don't know (and may be trying to find) is usually plotted on the vertical scale. In fact, the graph contains the same information regardless of which way round it is drawn!

Now let's take another example. Assume that the following measurements are made on an electronic component:

Temperature, t (°C)	10	20	30	40	50	60
Resistance, R (Ω)	105	110	115	120	125	130

The results of the experiment are shown plotted in graphical form in Fig. A6.8. Note that the graph consists of a straight line but that it does not pass through the *origin* of the graph (i.e. the point at which t and V are 0°C and 0 V respectively). The second most important feature to note (after having noticed that the graph is a straight line) is that, when $t = 0$°C, $R = 100 \, \Omega$.

Figure A6.6   Readings of current, i, and voltage, v, and can be used to construct the graph shown in Fig. A6.7

Figure A6.7   Graph of i plotted against v

Figure A6.8 Graph of resistance, R, plotted against temperature, t, for a thermistor

By looking at the graph we could suggest a relationship (i.e. an *equation*) that will allow us to find the resistance, R, of the component at any given temperature, t. The relationship is simply:

$$R = 100 + \frac{t}{2}\,\Omega$$

If you need to check that this works, just try inserting a few pairs of values from those given in the table. You should find that the equation balances every time!

Earlier we looked at equations. The shape of a graph is dictated by the equation that connects its two variables. For example, the general equation for a straight line takes the form:

$$y = mx + c$$

where y is the *dependent variable* (plotted on the vertical or *y-axis*), x is the *independent variable* (plotted on the horizontal or *x-axis*), m is the slope (or *gradient*) of the graph and c is the *intercept* on the *y-axis*. Figure A6.9 shows this information plotted on a graph.

The values of m (the gradient) and c (the y-axis intercept) are useful when quoting the specifications for electronic components. In the previous example, the electronic component (in this case a *thermistor*) has:

(a) a resistance of $100\,\Omega$ at $0°\text{C}$ (thus $c = 100\,\Omega$);
(b) a characteristic that exhibits an increase in resistance of $0.5\,\Omega$ per $°\text{C}$ (thus $m = 0.5\,\Omega/°\text{C}$).

Example 12

The following data was obtained whilst making measurements on an N-channel field effect transistor:

V_{GS} (V)	−1	−2	−3	−4	−5	−6
I_D (mA)	8	6	4	2	0.4	0.1

Plot a graph showing how drain current (I_D) varies with gate-source voltage (V_{GS}) and use the graph to determine:

(a) the value of I_D when $V_{GS} = 0\,\text{V}$; and
(b) the slope of the graph (expressed in mA/V) when $V_{GS} = -2\,\text{V}$.

The data has been shown plotted in Fig. A6.10. Note that, since the values of V_{GS} that we have been given are negative, we have plotted these to the left of the vertical axis rather than to the right of it.

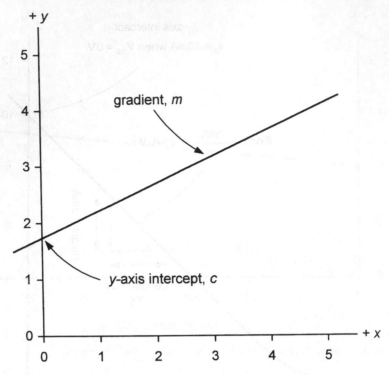

Figure A6.9   Definition of gradient (slope), m, and y-axis intercept, c

(a) We can use the graph to determine the crossing point (the *intercept*) on the drain current (I_D) axis. This occurs when $I_D = 10\,\text{mA}$.

(b) The slope (or *gradient*) of the graph is found by taking a small change in drain current and dividing it by a corresponding small change in gate-source voltage. In order to do this we have drawn a triangle on the graph at the point where $V_{GS} = -2\,\text{V}$. The slope of the graph is found by dividing the vertical height of the triangle (expressed in mA) by the horizontal length of the triangle (expressed in V). From Fig. A6.10 we can see that the vertical height of triangle is $2\,\text{mA}$ whilst its horizontal length is $1\,\text{V}$. The slope of the graph is thus given by $2\,\text{mA}/1\,\text{V} = \mathbf{2\,mA/V}$.

Square law graphs

Of course, not all graphs have a straight line shape. In the previous example we saw a graph that, whilst substantially linear, became distinctly curved at one end. Many graphs are curved rather than linear. One common type of curve is the square law. To put this into context, consider the relationship between the power developed in a load resistor and the

current applied to it. Assuming that the load has a resistance of $15\,\Omega$ we could easily construct a table showing corresponding values of power and current, as shown below:

Current, I (A)	0.25	0.5	0.75	1	1.25	1.5
Power, P (W)	0.94	3.75	8.44	15	23.44	33.75

We can plot this information on a graph showing power, P, on the vertical axis plotted against current, I, on the horizontal axis. In this case, P is the *dependent variable* and I is the *independent variable*. The graph is shown in Fig. A6.11.

It can be seen that the relationship between P and I in Fig. A6.11 is far from linear. The relationship is, in fact, a *square law relationship*. We can actually deduce this from what we know about the power dissipated in a circuit and the current flowing in the circuit. You may recall that:

$$P = I^2 R$$

where P represents power in watts, I is current in amps, and R is resistance in ohms.

Since R remains constant, we can deduce that:

$$P \propto I^2$$

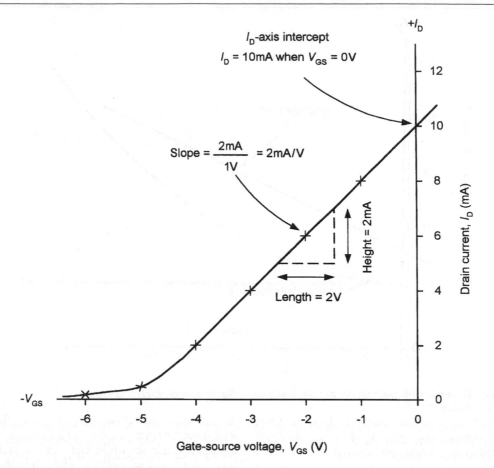

Figure A6.10   Drain current, I_D, plotted against gate-source voltage, V_{GS}, for a field effect transistor

In words, we would say that 'power is proportional to current squared'.

Many other examples of square law relationships are found in electronics.

More complex graphs

Many more complex graphs exist and Fig. A6.12 shows some of the most common types. Note that these graphs have all been plotted over the range $x = \pm 4$, $y = \pm 4$. Each graph consists of four quadrants. These are defined as follows (see Fig. A6.13):

First quadrant   Values of x and y are both positive

Second quadrant   Values of x are negative whilst those for y are positive

Third quadrant   Values of x and y are both negative

Fourth quadrant   Values of x are positive whilst those for y are negative

The straight line relationship, $y = x$, is shown in Fig. A6.12(a). This graph consists of a straight line with a gradient of 1 that passes through the *origin* (i.e. the point where $x = 0$ and $y = 0$). The graph has values in the first and third quadrants.

The relationship $y = x^2$ is shown in Fig. A6.12(b). This graph also passes through the origin but its gradient changes, becoming steeper for larger values of x. As you can see, the graph has values in the first and second quadrants.

The graph of $y = x^3$ is shown in Fig. A6.12(c). This *cubic law* graph is steeper than the square law of Fig. A6.12(b) and it has values in the first and third quadrants.

Figure A6.12(d) shows the graph of $y = x^4$. This graph is even steeper than those in Figs A6.12(b) and A6.12(c). Like the square law graph of

Figure A6.11   Power, P, plotted against current, I, showing a square law gaph

Fig. A6.12(b), this graph has values in the first and second quadrants.

The graph of $y = x^5$ is shown in Fig. A6.12(e). Like the cubic law graph of Fig. A6.12(c), this graph has values in the first and third quadrants.

Finally, Fig. A6.12(f) shows the graph of $y = 1/x$ (or $y = x^{-1}$). Note how the y values are very large for small values of x and very small for very large values of x. This graph has values in the first and third quadrants.

If you take a careful look at Fig. A6.12 you should notice that, for odd powers of x (i.e. x^1, x^3, x^5, and x^{-1}) the graph will have values in the first and third quadrant whilst for even powers of x (i.e. $8x^2$ and x^4) the graph will have values in the first and second quadrants.

Trigonometry

You might be wondering what trigonometry has to do with electronic servicing! Actually, a familiarity with some basic trigonometry is fundamental to developing an understanding of signal, waveforms and a.c. circuits. Indeed, the most fundamental waveform used in a.c. circuits is the sine wave and this wave is a trigonometric function.

Basic trigonometrical ratios

Trigonometrical ratios are to do with the way in which we measure angles. Take a look at the right-angled triangle shown in Fig. A6.14. This triangle has three sides: a, b and c. The angle that we are interested in (we have used the Greek symbol θ to denote this angle) is adjacent to side a and is opposite to side b. The third side of the triangle (the *hypotenuse*) is the longest side of the triangle.

In Fig. A6.14, the theorem of Pythagoras states that: 'the square on the hypotenuse is equal to the sum of the squares on the other two sides'. Writing this as an equation we arrive at:

$$c^2 = a^2 + b^2$$

where c is the hypotenuse and a and b are the other two sides.

Taking square roots of both sides of the equation we can see that:

$$c = \sqrt{a^2 + b^2}$$

Thus if we know two of the sides (for example, a and b) of a right-angled triangle we can easily find the third side (c).

The ratios $\dfrac{a}{c}$, $\dfrac{b}{c}$ and $\dfrac{a}{b}$ are known as the basic *trigonometric ratios*. They are known as sine (*sin*),

Figure A6.12   The first, second, third and fourth quadrants of a graph

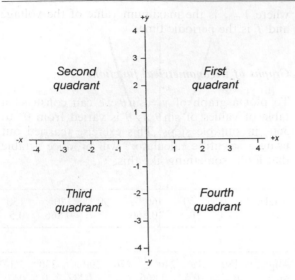

Figure A6.13 Some complex graphs and their equations

cosine (*cos*) and tangent (*tan*) of angle θ respectively. Thus:

$$\sin\theta = \frac{\text{opposite}}{\text{hypotenuse}} = \frac{a}{c}$$

and

$$\cos\theta = \frac{\text{adjacent}}{\text{hypotenuse}} = \frac{b}{c}$$

and

$$\tan\theta = \frac{\text{opposite}}{\text{adjacent}} = \frac{a}{b}$$

Trigonometrical equations

Equations that involve trigonometrical expressions are known as trigonometrical equations. Fortu-

nately they are not quite so difficult to understand as they sound! Consider the equation:

$$\sin\theta = 0.5$$

This equation can be solved quite easily using a calculator. However, before doing so, you need to be sure to select the correct mode for expressing angles on your calculator. If you are using a 'scientific calculator' you will find that you can set the angular mode to either *radian measure* or *degrees*. A little later we will explain the difference between these two angular measures but for the time being we shall use degrees.

If you solve the equation (by keying in 0.5 and pressing the *inverse sine* function keys) you should see the result 30° displayed on your calculator. Hence we can conclude that:

$$\sin 30° = 0.5$$

Actually, a number of other angles will give the same result! Try pressing the sine function key and entering the following angles in turn:

30°, 210°, 390° and 570°

They should all produce the same result, 0.5! This should suggest to you that the graph of the sine function repeats itself (i.e. the shape of the graph is *periodic*). In the next section we shall plot the sine function but, before we do we shall take a look at using radian measure to specify angles.

The *radian* is defined as the angle subtended by an arc of a circle equal in length to the radius of the circle. This relationship is illustrated in Fig. A6.15.

The circumference, *l*, of a circle is related to its radius, *r*, according to the formula:

$$l = 2\pi r$$

Thus

$$r = \frac{l}{2\pi}$$

Figure A6.14 A right-angled triangle

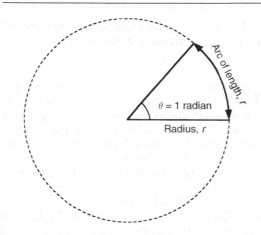

Figure A6.15   Definition of the radian

Now, since there are 360° in one complete revolution we can deduce that one radian is the same as $360°/2\pi = 57.3°$. On other words, to convert:

(a) degrees to radians multiply by 57.3;
(b) radians to degrees divide by 57.3.

It is important to note that one complete cycle of a periodic function (i.e. a waveform) occurs in a time, T. This is known as the *periodic time* or just the *period*. In a time interval equal to T, the angle will have changed by 360°. The relationship between time and angle expressed in degrees is thus:

$$\theta = \frac{T}{t} \times 360° \quad \text{and} \quad t = \frac{T}{\theta} \times 360°$$

Thus, if one complete cycle (360°) is completed in 0.02 s (i.e. $T = 20\,\text{ms}$) an angle of 180° will correspond to a time of 0.01 s (i.e. $t = 10\,\text{ms}$).

Conversely, if we wish to express angles in radians:

$$\theta = \frac{T}{t} \times 2\pi \quad \text{and} \quad t = \frac{T}{\theta} \times 2\pi$$

Thus, if one complete cycle (2π radians) is completed in 0.02 s (i.e. $T = 20\,\text{ms}$) an angle of π radians will correspond to a time of 0.01 (i.e. $t = 10\,\text{ms}$).

It should be apparent from this that, when considering waveforms, time and angle (whether expressed in degrees or radians) are interchangeable.

Note that the general equation for a sine wave *voltage* is:

$$v = V_{\text{max}} \sin(2\pi f t) = V_{\text{max}} \sin\left(6.28 \times \frac{t}{T}\right)$$

where V_{max} is the maximum value of the voltage and T is the periodic time.

Graphs of trigonometrical functions

To plot a graph of $y = \sin\theta$ we can construct a table of values of $\sin\theta$ as θ is varied from 0° to 360° in suitable steps. This exercise (carried out using a scientific calculator) will produce a table that looks something like this:

Angle, θ	0°	30°	60°	90°	120°	150°
$\sin\theta$	0	0.5	0.866	1	0.866	0.5

Angle, θ	180°	210°	240°	270°	300°	330°	360°
$\sin\theta$	0	−0.5	−0.866	−1	−0.866	−0.5	0

Plotting the values in the table reveals the graph shown in Fig. A6.16.

We can use the same technique to produce graphs of $\cos\theta$ and $\tan\theta$, as shown in Figs A6.17 and A6.18.

Example 13

The voltage applied to a high voltage rectifier is given by the equation:

$$v = 800 \sin(314t)$$

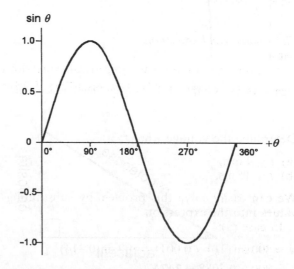

Figure A6.16   Graph of the sine function, $y = \sin\theta$

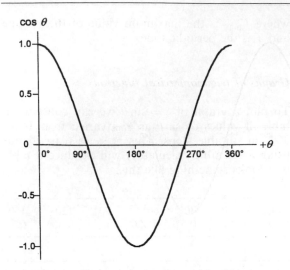

Figure A6.17   Graph of the cosine function, $y = \cos\theta$

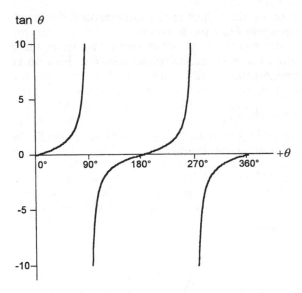

Figure A6.18   Graph of the tangent function, $y = \tan\theta$

Determine the voltage when:

(a) $t = 1\,\text{ms}$;
(b) $t = 15\,\text{ms}$.

We can easily solve this problem by substituting values into the expression.

In case (a):

$$v = 800\sin(314 \times 0.001) = 800\sin(0.314)$$

$$= 800 \times 0.3088 = \mathbf{247\,V}$$

In case (b):

$$v = 800\sin(314 \times 0.015) = 800\sin(4.71)$$

$$= 800 \times -1 = \mathbf{-800\,V}$$

Example 14

The voltage appearing at the secondary winding of a transformer is shown in Fig. A6.19. Determine:

(a) the period of the voltage;
(b) the value of the voltage when (i) $t = 5\,\text{ms}$ and (ii) $t = 35\,\text{ms}$;
(c) the maximum and minimum values of voltage;
(d) the average value of the voltage during the time interval 0 to 40 ms;
(e) the equation for the voltage.

All of the above can be obtained from the graph of Fig. A6.19, as follows:

(a) period, $T = \mathbf{20\,ms}$ (this is the time for one complete cycle of the voltage);
(b) at $t = 5\,\text{ms}$, $v = \mathbf{20\,V}$; at $t = 35\,\text{ms}$, $v = \mathbf{-20\,V}$;
(c) $v_{\text{max}} = \mathbf{+20\,V}$ and $v_{\text{min}} = \mathbf{-20\,V}$;
(d) $v_{\text{av.}} = \mathbf{0\,V}$ (the positive and negative cycles of the voltage are equal);
(e) $v = \mathbf{20\sin 314\mathit{t}}$ (the maximum value of the voltage, V_{max}, is 20 V and $T = 0.02\,\text{s}$).

Boolean algebra

Boolean algebra is the algebra that we use to describe and simplify logic expressions. An understanding of Boolean algebra can be extremely useful when troubleshooting any circuit that employs digital logic. It is, for example, very useful to be able to predict the output of a logic circuit when a particular input condition is present.

Boolean notation

Just as with conventional algebra, we use letters to represent the variables in these expressions. The symbols used in Boolean algebra do not, however, have the same meanings as they have in conventional algebra:

Symbol	Meaning
.	Logical AND
+	Logical OR
⊕	Logical exclusive-OR
−	Logical NOT

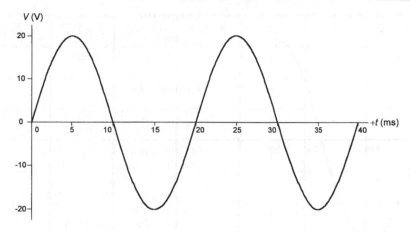

Figure A6.19   See Example 14

Thus '*A* AND *B*' can be written ($A \cdot B$) whereas '*A* OR *B*' is written ($A + B$). The bar symbol, $\overline{}$, used above a variable (or over a part of an expression) indicates inversion (the NOT function). Thus 'NOT *A*' is written $\overline{A}$ and 'NOT (*A* AND *B*)' is written $\overline{A \cdot B}$. Note that $\overline{A \cdot B}$ is *not* the same as $\overline{A} \cdot \overline{B}$.

It is important to note that a logic variable (such as *A* or *B*) can exist in only two states variously (and interchangeably) described as true or false, asserted or non-asserted, 1 or 0, etc. The two states are mutually exclusive and there are no 'in between' conditions.

Truth tables

We often use tables to describe logic functions. These tables show all possible logical combinations and they are called *truth tables*. The columns in a truth table correspond to the input variables and an extra column is used to show the resulting output. The rows of the truth tables show all possible states of the input variables. Thus, if there are two input variables (*A* and *B*) there will be four possible input states, as follows:

$A = 0, B = 0$
$A = 0, B = 1$
$A = 1, B = 0$
$A = 1, B = 1$

Basic logic functions

The basic logic functions are: OR, AND, NOR and NAND. We can describe each of these basic logic functions using truth tables as shown in Fig. A6.20 (for two input variables) and in Fig. A6.21 (for three input variables).

In general, a logic function with *n* input variables will have 2^n different possible states for those inputs. Thus, with two input variables there will be four possible input states, for three input variables there will be eight possible states, and for four input variables there will be 16 possible states.

Take a look at Fig. A6.20 and Fig. A6.21 and compare the output states for AND with those for NAND as well as OR with those for NOR. In every case, you should notice that the output states are complementary – in other words, all of 0s in the AND column have been inverted to become 1s in the NAND column, and vice versa. The same rule also applies to the OR and NOR columns.

Venn diagrams

An alternative technique to that of using a truth table to describe a logic function is that of using a Venn diagram. This diagram consists of a number

A	B	A + B	A·B	$\overline{A + B}$	$\overline{A \cdot B}$
0	0	0	0	1	1
0	1	1	0	0	1
1	0	1	0	0	1
1	1	1	1	0	0

Figure A6.20   Truth tables for the basic logic functions using two input variables: *A* and *B*

A	B	C	A+B+C	A·B·C	$\overline{A+B+C}$	$\overline{A·B·C}$
0	0	0	0	0	1	1
0	0	1	1	0	0	1
0	1	0	1	0	0	1
0	1	1	1	0	0	1
1	0	0	1	0	0	1
1	0	1	1	0	0	1
1	1	0	1	0	0	1
1	1	1	1	1	0	0

Figure A6.21   Truth tables for the basic logic functions using three variables: *A*, *B* and *C*

of overlapping areas that represent all of the possible logic states.

A simple Venn diagram with only one variable is shown in Fig. A6.22. The shaded area represents the condition when variable *A* is *true* (i.e. logic 1). The non-shaded area represents the condition when variable *A* is *false* (i.e. logic 0). Obviously, variable *A* cannot be both true and false at the same time hence the shaded and non-shaded areas are mutually exclusive.

Figure A6.23 shows Venn diagrams for the basic logic functions: AND, OR, NAND and NOR. You should note that these diagrams convey exactly the same information as the truth tables in Fig. A6.20 but presented in a different form.

Karnaugh maps

A Karnaugh map consists of a square or rectangular array of cells into which 0s and 1s may be placed to indicate false and true respectively. Two alternative representations of a Karnaugh map for two variables are shown in Fig. A6.24.

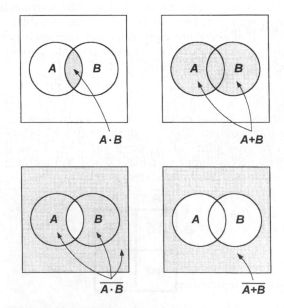

Figure A6.23   Venn diagrams for the basic logic functions

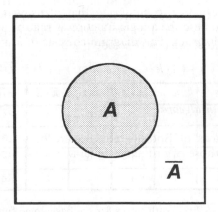

Figure A6.22   Venn diagram for a single logic variable, *A*

Figure A6.24   Alternative representations for a Karnaugh map showing two variables

The relationship between Boolean logic expressions for two variables and their Karnaugh maps is illustrated in Fig. A6.25 whilst Fig. A6.26 shows how the basic logic functions, AND, OR, NAND and NOR, can be plotted on Karnaugh maps.

Adjacent cells within a Karnaugh map may be grouped together in rectangles of two, four, eight, etc. cells in order to simplify a complex logic expression. Taking the NAND function, for example, the two groups of two adjacent cells in the Karnaugh map correspond to $\overline{A}$ or $\overline{B}$, as shown in Fig. A6.27.

We thus conclude that:

$$\overline{A \cdot B} = \overline{A} + \overline{B}$$

Once again note the overscore characters are used to indicate the inverting function.

This important relationship is known as *De Morgan's theorem.*

Karnaugh maps can be drawn showing two, three, four, or even more variables. The Karnaugh map shown in Fig. A6.28 is for three variables – note that there are eight cells and that each cell corresponds to one of the eight possible combinations of the three variables: A, B and C.

The technique of grouping cells together is an extremely powerful one. On a Karnaugh map showing three variables, if two adjacent cells both contain 1s they can be grouped together to produce an expression containing just two variables. Similarly, if four adjacent cells can be

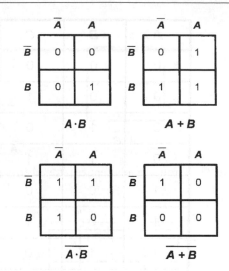

Figure A6.26   Karnaugh maps for the basic logic functions

grouped together, they reduce to an expression containing only one variable.

It is also important to note that the map is a continuous surface which links edge to edge. This allows cells at opposite extremes of a row or column to be linked together. The four corner cells of a four (or more) variable map may likewise be grouped together (provided they all contain 1s).

As an example, consider the following Boolean function:

$$(A \cdot B \cdot C) + (A \cdot B \cdot \overline{C}) + (A \cdot \overline{B} \cdot C) + (A \cdot \overline{B} \cdot \overline{C})$$

Figure A6.29 shows how this function is plotted on the Karnaugh map.

As before, we have placed a 1 in each cell. Then we can begin to link together adjacent cells – notice how we have reduced the diagram to a group of four cells equivalent to the logic variable, A. In other words, the value of variables B and C, whether 0 or 1, will have no effect on the logic function. From Fig. A6.30 we can conclude that:

$$(A \cdot B \cdot C) + (A \cdot B \cdot \overline{C}) + (A \cdot \overline{B} \cdot C) + (A \cdot \overline{B} \cdot \overline{C}) = A$$

Boolean algebra

The laws of Boolean algebra are quite different from those of ordinary algebra. For example:

$$A \cdot 0 = 0$$
$$A \cdot 1 = A$$
$$A + 0 = A$$
$$A + 1 = 1$$

	$\overline{A}$	A
$\overline{B}$	$\overline{A} \cdot \overline{B}$	$A \cdot \overline{B}$
B	$\overline{A} \cdot B$	$A \cdot B$

Figure A6.25   Karnaugh map for two variables showing the Boolean logic expressions for each cell

Figure A6.27 Grouping adjacent cells together to prove De Morgan's theorem

Figure A6.28 Karnaugh map for three variables

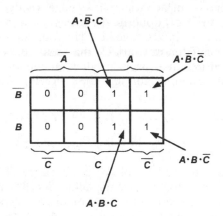

Figure A6.29 Plotting terms on a three-variable Karnaugh map

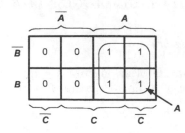

Figure A6.30 Grouping adjacent cells together to simplify a complex Boolean expression

The *Commutative Law* shows us that:

$$A + B = B + A$$

and also that:

$$A \cdot B = B \cdot A$$

The *Associative Law* shows us that:

$$A + (B + C) = (A + B) + C = A + B + C$$

and also that:

$$A \cdot (B \cdot C) = (A \cdot B) \cdot C = A \cdot B \cdot C$$

The *Distributive Law* shows us that:

$$A \cdot (B + C) = A \cdot B + B \cdot C$$

and also that:

$$A + (B \cdot C) = (A + B) \cdot (A + C)$$

All of the above laws can be proved using the various techniques that we met earlier.

Boolean simplification

Using the laws of Boolean algebra and De Morgan's theorem, $\overline{A \cdot B} = \overline{A} + \overline{B}$ (see earlier), we can reduce complex logical expressions in order to minimize the number of variables and the number of terms. For example, the expression that we met in B33.6.4 can be simplified as follows:

$$(A \cdot B \cdot C) + (A \cdot B \cdot \overline{C}) + (A \cdot \overline{B} \cdot C) + (A \cdot \overline{B} \cdot \overline{C})$$
$$= (A \cdot B) + (A \cdot \overline{B}) = A$$

Here we have eliminated the variable C from the first and second pair of terms and then eliminated B from the result. You might like to compare this with the Karnaugh map that we drew earlier!

Example 15

Part of the control logic for an intruder alarm is shown in Fig. A6.31. Determine the logic required to select the Y4 output and verify that the Y3 output is permanently selected and unaffected by the states of S1 and S2.

Each logic gate shown in Fig. A6.31 is a two input NAND gate. We can construct the truth table for the control circuit (see Fig. A6.32) by placing all four possible combinations of 0 and 1 on the S1 and S2 inputs and tracing the logic states through the circuit. We have shown the first stage of this process (the logic conditions that produce the first line of the truth table) in Fig. A6.32.

From Fig. A6.32 it can be seen that the Y4 output corresponds to the AND logic function:

$$Y4 = S1 \cdot S2$$

You can also see from Fig. A6.32 that the Y3 output remains at logic 1 regardless of the states of S1 and S2.

Logarithms

Many of the numbers that we have to deal with in electronics are extremely large whilst others can be

Figure A6.31   Intruder alarm logic – see Example 15

S1	S2	Y1	Y2	Y3	Y4
0	0	1	1	1	0
0	1	1	1	1	0
1	0	1	0	1	0
1	1	0	1	1	1

Logic states corresponding to this line are shown in Fig. **A6.31**

Figure A6.32   Truth table for the control logic shown in Fig. A6.31

extremely small. The problem of coping with a wide range of numbers is greatly simplified by using logarithms. Because of this, we frequently employ logarithmic scales, bels (B) and decibels (dB), when comparing values. Note that a decibel (dB) is simply one-tenth of a bel (B). Hence:

$$10dB = 1B$$

It is important to note that the bel is derived from the logarithm of a ratio – it is not a unit like the volt (V) or ampere (A).

The logarithm (to the base 10) of a number is defined as the power that 10 must be raised to in order to equal that number. This may sound difficult but it isn't, as a few examples will show:

Because $10 = 10^1$, the logarithm (to the base 10) of 10 is 1.
Because $100 = 10^2$, the logarithm (to the base 10) of 100 is 2.
Because $1000 = 10^3$, the logarithm (to the base 10) of 1000 is 3 and so on.

Writing this as a formula we can say that:

If $x = 10^n$, the logarithm (to the base 10) of x is n.

You might like to check that this works using the examples that we just looked at!

You may be wondering why we keep writing 'to the base 10'. Logarithms can be to *any* base but 10 is perhaps the most obvious and useful.

Logarithmic notation

The notation that we use with logarithms is quite straightforward:

If $x = 10^n$ we say that $\log_{10}(x) = n$.

The subscript that appears after the 'log' reminds us that we are using base 10 rather than any other (unspecified) number. Rewriting what we said earlier using our logarithmic notation gives:

$$\log_{10}(10) = \log_{10}(10^1) = 1$$

$$\log_{10}(100) = \log_{10}(10^2) = 2$$

$$\log_{10}(1000) = \log_{10}(10^3) = 3$$

and so on.

So far we have used values (10, 100 and 1000) in our examples that just happen to be exact powers of 10. So how do we cope with a number that isn't an exact power of 10? If you have a scientific calculator this is quite easy! Simply enter the number and press the 'log' button to find its logarithm (the instructions provided with your calculator will tell you how to do this). You might like to practice with a few values, for example:

$\log_{10}(56) = 1.748$

$\log_{10}(1.35) = 0.1303$

$\log_{10}(4028) = 3.605$

$\log_{10}(195\,891) = 5.292$

$\log_{10}(0.3175) = -0.4983$

Finally, there's another advantage of using logarithmic units (i.e. bels and decibels) in electronics. If two values are multiplied together the result is equivalent to adding their values expressed in logarithmic units. To put this into context, let's assume that the following items are connected in the signal path of an amplifier (see Fig. A6.33):

Pre-amplifier; input power = 1 mW, output power = 80 mW
Tone control; input power = 80 mW, output power = 20 mW
Power amplifier; input power = 20 mW, output power = 4 W

Component	Input power	Output power	Power gain (expressed as a ratio)	Power gain (expressed in decibels)
Pre-amplifier	1 mW	80 mW	80	19
Tone control	80 mW	20 mW	0.25	−6
Power amplifier	20 mW	4 W	200	23

The overall power gain can be calculated from the following:

Overall power gain (expressed as a ratio)

$$= \frac{\text{output power}}{\text{input power}} = \frac{4\,\text{W}}{1\,\text{mW}} = 4000$$

Converting this gain to decibels gives $10 \log_{10}(4000) = 36\,\text{dB}$.

Adding the power gains from the right-hand column gives: $19 - 6 + 23 = 36\,\text{dB}$

Example 16

An amplifier has a power gain of 27 000. Express this in bels (B) and decibels (dB).

In bels, power gain $= \log_{10}(27\,000) = \mathbf{4.431\,B}$

In decibels, power gain $= 10 \times \log_{10}(27\,000) = 10 \times 4.431 = \mathbf{44.31\,dB}$

Example 17

An amplifier has a power gain of 33 dB. What input power will be required to produce an output power of 11 W?

Now power gain expressed in decibels will be given by:

$$10 \log_{10}\left(\frac{P_{\text{out}}}{P_{\text{in}}}\right)$$

Thus $33 = 10 \log_{10}\left(\frac{11\,\text{W}}{P_{\text{in}}}\right)$

Dividing both sides by 10 gives:

$$3.3 = \log_{10}\left(\frac{11\,\text{W}}{P_{\text{in}}}\right)$$

Taking the antilog of both sides gives:

$$\text{antilog}(3.3) = \frac{11\,\text{W}}{P_{\text{in}}}$$

$$1995 = \frac{11\,\text{W}}{P_{\text{in}}}$$

Figure A6.33   Amplifier system showing power gains and losses from which:

$$P_{\text{in}} = \frac{11\,\text{W}}{1995} = 5.5\,\text{mW}$$

Note that the 'antilog' or 'inverse log' function can be found using your calculator. This sometimes appears as a button marked '10^x' or is obtained by pressing 'shift' (to enable the inverse function) and then 'log'. In any event, you should refer to your calculator's instruction book for more information!

Exponential growth and decay

In electronics we are often concerned with how quantities grow and decay in response to a sudden change. In such cases the laws that apply are the same as those that govern any natural system. The formula that relates to *exponential growth* takes the form:

$$y = Y_{\text{max}}(1 - e^x)$$

whilst that which relates to *exponential decay* takes the form:

$$y = Y_{\text{max}}e^{-x}$$

In the case of exponential growth, the value, Y_{max}, is simply the maximum value for y. In other words it is approximately the value that y will reach after a very long time (note that, even though y gets very close in value to it, Y_{max} is never quite reached).

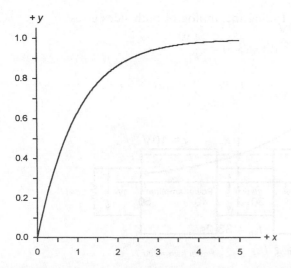

Figure A6.34 A graph illustrating exponential growth

In the case of exponential decay, the value, Y_{max}, is the initial value of y. In other words it is the value of y before the decay starts. After a very long period of time, the value of y will reach approximately zero (note that, even though y gets very close in value to zero, it is never quite reached).

Figure A6.34 shows a graph showing exponential growth whilst Fig. A6.35 shows a graph showing exponential decay. In both cases Y_{max} has been set to 1. The name given to the constant, e, is the exponential constant and it has a value of approximately 2.713. The value of the constant is normally accessible from your calculator (in the same way that the value of π is stored as a constant in your calculator).

Example 18

A capacitor is charged to a potential of 100 V. It is then disconnected from its charging source and left to discharge through a resistor. If it takes 10 s for the voltage to fall to 50 V, how long will it take to fall to 10 V?

Here we are dealing with exponential decay. If we use y to represent the capacitor voltage at any instant, and since the initial voltage is 100 V, we can determine the value of x (i.e. x_1) when the voltage reaches 50 V by substituting into the formula:

$$y = Y_{\text{max}}e^{-x}$$

hence:

$$50 = 100e^{-x_1}$$

from which:

$$e^{-x_1} = \frac{50}{100} = 0.5$$

taking logs (to the base e this time) of both sides:

$$\log_e(e^{-x_1}) = \log_e(0.5)$$

thus:

$$-x_1 = -0.693$$

hence:

$$x_1 = 0.693$$

We can use a similar process to find the value of x (i.e. x_2) that corresponds to a capacitor voltage of 10 V:

$$y = Y_{\text{max}}e^{-x}$$

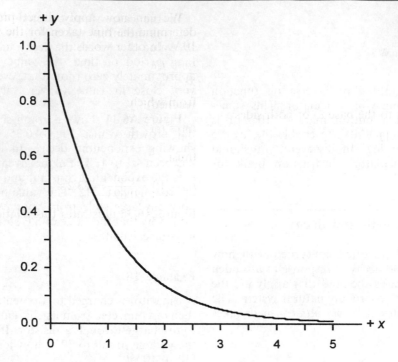

Figure A6.35   A graph illustrating exponential decay

Figure A6.36   Capacitor voltage on discharge – see Example 18

hence:

$10 = 100e^{-x_2}$

from which:

$e^{-x_2} = \dfrac{10}{100} = 0.1$

taking logs (again to the base e) of both sides:

$\log_e(e^{-x_2}) = \log_e(0.1)$

thus:

$-x_2 = -2.3$

hence:

$x_2 = 2.3$

We can now apply direct proportionality to determine the time taken for the voltage to reach 10 V. In other words:

$\dfrac{t_{50}}{x_1} = \dfrac{t_{10}}{x_2}$

from which:

$t_{10} = t_{50} \times \dfrac{x_2}{x_1}$

thus:

$t_{10} = 10 \times \dfrac{2.3}{0.693} = \mathbf{33.2\,s}$

Figure A6.36 illustrates this relationship.

Appendix 7
Useful web links

Semiconductor manufacturers

Analog Devices	http://www.analog.com/
Burr-Brown	http://www.burr-brown.com/
Dallas Semiconductor	http://www.dalsemi.com/
Elantec	http://www.elantec.com/
Fujitsu	http://www.fujitsu-fme.com/
Harris Semiconductor	http://www.harris.com/
Holtek Semiconductor	http://www.holtek.com/default.htm
Intel	http://www.intel.com/
International Rectifier	http://www.irf.com/
Linear Technology	http://www.linear-tech.com/
Maxim	http://www.maxim-ic.com/
Microchip Technology	http://www.microchip.com/
National Semiconductor	http://www.nsc.com/
Philips	http://www.semiconductors.philips.com/
Samsung	http://www.samsung.com/
Seiko ECD	http://www.seiko-usa-ecd.com/
SGS-Thompson	http://eu.st.com/stonline/index.htm
Siemens	http://www.siemens.de/
Sony	http://www.sony.com/
TDK Semiconductor	http://www.tsc.tdk.com/
Texas Instruments	http://www.ti.com/
Zilog	http://www.zilog.com/

Electronic component suppliers

Farnell Electronic Components	http://www.farnell.co.uk/
Greenweld	http://www.greenweld.co.uk/
Magenta	http://www.magenta2000.co.uk/
Maplin Electronics	http://www.maplin.co.uk/
Quasar Electronics	http://www.quasarelectronics.com/home.htm
RS Components	http://www.rswww.com/

Magazines

Everyday with Practical Electronics	http://www.epemag.wimborne.co.uk/
EPE Online	http://www.epemag.com/mainhom.htm

Clubs and societies

American Radio Relay league	http://www.arrl.org/
British Amateur Electronics Club	http://members.tripod.com/baec/
Radio Society of Great Britain	http://www.rsgb.org.uk/

Multimedia electronics CD-ROMs

Matrix Multimedia	http://www.matrixmultimedia.co.uk/

Miscellaneous

WebEE Electronic Engineering Homepage   http://www.web-ee.com/

Index